NCAER

INDIA POLICY FORUM 2011|12

VOLUME 8

EDITED BY

Shekhar Shah

Barry Bosworth

Arvind Panagariya

NATIONAL COUNCIL OF APPLIED ECONOMIC RESEARCH
New Delhi

BROOKINGS INSTITUTION
Washington, D.C.

SAGE www.sagepublications.com
Los Angeles • London • New Delhi • Singapore • Washington DC

First published in 2012 by

SAGE Publications India Pvt Ltd
B1/I-1 Mohan Cooperative Industrial Area
Mathura Road, New Delhi 110 044, India
www.sagepub.in

SAGE Publications Inc
2455 Teller Road
Thousand Oaks, California 91320, USA

SAGE Publications Ltd
1 Oliver's Yard, 55 City Road
London EC1Y 1SP, United Kingdom

SAGE Publications Asia-Pacific Pte Ltd
33 Pekin Street
#02-01 Far East Square
Singapore 048763

Library of Congress Serial Publication Data applied for

ISSN: 0973-4805
ISBN: 978-81-321-0973-0 (PB)

Published by Vivek Mehra for SAGE Publications India Pvt Ltd, typeset in 10.5/13 pt Times by Star Compugraphics Private Limited, Delhi and printed at Chaman Enterprises, New Delhi.

NCAER

INDIA POLICY FORUM 2011|12

VOLUME 8

PURPOSE

India Policy Forum 2011–12 comprises papers and highlights of the discussions at the eighth India Policy Forum (IPF) held in New Delhi on July 12–13, 2011. The IPF is a joint venture of the National Council of Applied Economic Research (NCAER) in New Delhi and the Brookings Institution in Washington, D.C. The IPF explores India's rapidly evolving—and sometimes tumultuous—economic transition and its underlying policy frameworks and reforms using policy-relevant, empirical research.

An international Research Panel of India-based and overseas researchers with an abiding interest in India supports this initiative through advice, active participation at the IPF, and the search for innovative papers that promise fresh insights. An international Advisory Panel of distinguished economists provides overall guidance. Members of the two IPF panels are listed below.

Papers appear in this publication after detailed revisions based on discussants' comments at the IPF and the guidance provided by the editors after the IPF. To allow readers to get a sense of the richness of the conversations that happen at the IPF, discussants' comments are also included here, as is a summary of the general discussion on each paper. The papers represent the views of the individual authors and do not imply any agreement by those attending the conference, those providing financial support, or the officers and staff of NCAER or Brookings.

ADVISORY PANEL

RESEARCH PANEL

PARTNERS

NCAER and Brookings gratefully acknowledge the generous financial support for IPF 2011 from the State Bank of India, HDFC Ltd., Reliance Industries Ltd., Citibank NA, and IDFC Ltd.

The support reflects the deep commitment of these organizations and their leadership to rigorous policy research that helps promote informed policy debates and sound, evidence-based policymaking. Almost all these funders have been with the IPF since its inception, so their support also reflects their continuing confidence in the India Policy Forum to promote such research and open debate in India and elsewhere.

CORRESPONDENCE

Correspondence about papers in this Volume should be addressed directly to the authors (each paper contains the email address of the corresponding author). Manuscripts are not accepted for review because this Volume is devoted exclusively to invited contributions. Feedback on the IPF Volume itself may be sent to: The Editor, India Policy Forum, NCAER, 11, Indraprastha Estate, New Delhi 110 002 or to ipf@ncaer.org.

NCAER TEAM

Geetu Makhija
Sangita Chaudhary
Shikha Vasudeva
Sarita Sharma
Jagbir Singh Punia
P.P. Joshi
Sudesh Bala

Editors' Summary

The eighth conference of the India Policy Forum was held on July 12–13, 2011. This issue contains the final versions of the papers and discussions presented at the conference. The papers cover a range of important policy issues including bank regulation in the aftermath of the global financial crisis, controls on external capital flows, access to health care, and some implications of rising global food prices. In addition to the working sessions of the conference, Montek S. Ahluwalia, Deputy Chairman of the Indian Planning Commission, delivered a public lecture on the challenges faced under the 12th five-year plan. The IPF conference concluded with a lively Policy Roundtable on "From Price Distortions to Subsidies that Work for the Poor."

The global financial crisis, which began in the fall of 2007 and progressively worsened in subsequent years, hit the Indian economy in 2008. Equity market prices fell by nearly 60 percent in the first 10 months of the year. Foreign capital outflows placed pressure on Indian corporations and banking institutions, leading to a lessening of liquidity and credit. The pressures eventually spilled over into the real economy and the global slowdown also resulted in a slump in demand for exports. GDP growth slowed from 9 percent in 2007 to 6 percent in 2008. Eventually, the Government of India announced wide-ranging stimulus packages in 2009 that appeared to restore the economy back to its pre-2008 growth rate.

An important observation during 2008–09 was the apparent weakness of private financial firms against the (relative) growing strength of public sector or state-owned banks. Historically, Indian banks had been wholly owned by the government. In the 1990s, the government reduced its stake and allowed private banks and foreign players to enter the market, but the Indian financial system retains a substantive public sector ownership. During the crisis, the public sector banks appeared to fare much better than the private ones. Public sector bank deposits grew by 24.1 percent in 2009, as compared to 22.9 percent in 2008, while private sector deposit growth slowed from 19.9 percent to 8 percent for the same period. This apparent strength of public sector banks has raised the issue of whether it might be better to slow down the privatization of the Indian banking system.

The paper by Viral Acharya examines the relative performance of private and public sector banks. Contrary to the above impression, he argues that public banks were riskier than private sector ones, but their sovereign backing allowed their deposit base to grow in spite of their riskiness, and it actually undermined the performance of the private sector banks. Based on standard measures of systemic risks, public sector banks were actually riskier than their private counterparts. The marginal expected shortfall (MES) measure, used to calculate the systemic risk of financial institutions, was higher on average in 2007 for public sector banks (4.34 percent) compared to private sector banks (3.58 percent), indicating that public sector banks were riskier. During the crisis, however, public sector banks performed better than private banks due to a shift of deposits from private to public sector banks. While private sector banks with high exposure to the crisis experienced substantial deposit contraction during the crisis, counter-intuitively, within the public sector greater vulnerability to a crisis, in fact, led to greater deposit expansion.

Acharya argues that the phenomenon can be explained by the government backing of public sector banks. Perceptions of a greater likelihood that the more vulnerable public sector banks would be bailed out in the event of a failure led to a shift of investor funds and deposits. He shows that as part of the stimulus programs, public sector banks were promised capital injections to maintain their capital ratios. In addition, the generous backing of the Indian government allowed public sector banks to come out in the retail sector with inexpensive housing, auto, and education loans, which instead of drawing in new customers to stimulate the economy actually undercut competitors. Therefore, government measures may have helped bolster public sector banks, but they also made it difficult for private financial firms to compete with them.

The idea of financial firms with access to government guarantees faring better during the crisis is not a case just specific to India, but was seen in other countries as well. A specific example, emphasized by Acharya, was the growth of the government-sponsored enterprises (GSEs)—Fannie Mae and Freddie Mac—in the United States, which had explicit government support and ready access to central bank emergency lending.

The paper points to several parallels between the Indian government's control over financial institutions and the role of Fannie Mae and Freddie Mac in the United States. First, affordable home ownership goals in the United States are similar to "priority sector lending norms" in India. Second, the GSEs in the United States underwrote over 50 percent of mortgage credit risk in the residential housing market, corresponding to the large share of state-owned banks in India. Third, the forbearance exhibited toward the GSEs closely mirrors the reluctance of the Indian government to shut down any state-owned banks. Fourth, the GSEs crowded out the private financial sector in the prime residential housing market and eventually also in affordable housing loans.

With so many similarities between the two sets of institutions, Acharya argues that it is important to consider whether some of the same negative consequences experienced in the United States case might occur in India as well. There was a "race to the bottom" in finance between the public sector and the private sector in the United States. Also, in the aftermath of the crisis in the United States, the poor were especially affected, and the same could be true in India where the bulk of the priority-sector lending goes to farming loans, but there is no quality control of the loans because these state-owned banks never fail. Indian policymakers should use this and other examples as a note of caution against drawing conclusions favoring a state presence in the financial sector.

Examining the performance of state-owned banks in a systemic crisis relative to private sector banks with lesser access to government guarantees is not a sound basis for assessing the overall attractiveness of state presence in the financial sector. The growth of state-owned banks in the recent crisis thus raises several questions. How will the private-sector banks in India respond to the emerging strength of public sector banks? Can India afford such a race to the bottom? Would it not be better to create a level playing field where all financial firms are private, but are charged upfront for government guarantees in the form of deposit insurance? The Indian state-owned banks are right now in the best of health. Since many of them are already publicly listed, the way to their privatization and creation of a level playing field in the financial sector is relatively straightforward and needs to be undertaken without any further adieu.

The paper by Ajay Mahal and Victoria Fan, focuses on the challenge of reforming the health-care system in India. Accelerated economic growth has brought about a significant increase in public revenues in the last three decades. On the one hand, this has allowed the government to expand the social programs, but on the other, it has also led to increased demands for the services that the programs provide. The right to employment in rural areas and the right to elementary education are already in place in the form of the Mahatma Gandhi National Rural Employment Guarantee Scheme (MGNREGS) and the Right to Education Act of 2009, respectively. The Cabinet has also approved a right to food bill, and it awaits a vote in the Parliament. The pressure is next building for a right to health bill that would provide universal health coverage. The ensuing debate is taking place in the context of the National Rural Health Mission, which was launched in 2005, the Rashtriya Swasthya Bima Yojana, and various state-level health insurance programs. A recent High Level Expert Group (HLEG), appointed by the Planning Commission, has called for universal health coverage (UHC) with the government footing the bill for all.

Against this background, Mahal and Fan provide a comprehensive discussion of three aspects of health reform in India. First, what should the UHC cover and what should be the role of the state in implementing it? Second, what are the main challenges in expanding the health coverage in India, and should reform attempt to achieve UHC or something less ambitious? Finally, what are the financial implications of expanded coverage?

The authors begin by noting that government intervention in the health sector can be justified on both efficiency and equity grounds. Interventions aimed at prevention, health promotion, health insurance, and regulatory oversight help to improve efficiency while the support to the poor for inpatient care and, within limits, outpatient curative care are justified on equity grounds. Provisions for disadvantaged groups, including children, can also be defended as improving efficiency.

The authors point to two common mechanisms to combine insurance with assistance to the poor. The first mechanism uses separate pools of funds for the nonpoor and poor. The poor would be subsidized from general revenues, and an additional publicly subsidized fund would be earmarked for child health services and preventive and health promotion services. The nonpoor would pay for their benefits by way of premiums for regulated insurance. Alternatively, if they are left to their own devices, they would pay out of pocket for health services or purchase private insurance when available, although this is sometimes inefficient. The second mechanism involves a common fund pool for everyone and may involve the rich paying for their benefits by way of additional premiums. Again, separate funds could be allocated for elements of a public goods nature or with significant externalities, such as preventive care and health promotion activities.

Both of the approaches can, in principle, lead to UHC defined as access to a minimum set of services that are affordable and physically accessible. Yet, when access to a minimum set of publicly funded services is defined as a legally enforceable "right," problems of equity will arise if a country is not wealthy enough to support a wide range of services for the full population. In a resource-constrained environment, even if services are in principle accessible to all for free, better-off sections of society will be more adept at obtaining the benefits.

The authors go on to analyze the challenges facing the implementation of expanded coverage including defining the benefits to be covered by public subsidies, the groups to be targeted, the management of the funds used to provide health services, and the methods used to pay health-care providers. In deciding priority interventions, policymakers could adopt a strategy of prioritizing those interventions for which the health returns are the highest per dollar, which usually means a focus on prevention and promotion programs. From India's perspective, this means that considerable research into the costs and cost-effectiveness of interventions will be needed to support policy on an ongoing basis.

Key gaps in India are in the ability to design payments for health services in ways that serve the objectives of good quality, especially performance-linked payment systems. The managers of the fund may themselves need incentives to carry out their responsibilities effectively, necessitating some form of competition for the "right to manage" and community oversight. How providers are paid can have major implications for the efficiency with which resources are used and the quality of services. Choosing among the different options involves a number of trade-offs. Fee-for-service systems that are the predominant means of paying for health care in India are less likely to have adverse impacts on quality of health services, but they create incentives for excessive and inefficient use of resources. Prospective payment systems, including capitation, can contain costs but could harm the quality of health services received. Prospective payments that are risk-adjusted or combined with performance incentives can potentially achieve a better balance between the goals of quality and efficiency, but they have extensive information requirements.

Transfers to consumers offer another avenue to influence health service use. There is some evidence on the effectiveness of conditional transfers in influencing prevention behavior, both from India (under the JSY program) and elsewhere. Unconditional transfers to poor households in lieu of outpatient care coverage have also been proposed, but evidence on their effectiveness is limited.

The state of the supply—the provision of health services—is a major implementation challenge in the Indian health sector. Both the HLEG report and the National Commission on Macroeconomics and Health (NCMH) report preceding it highlight the need to upgrade health services in the public sector, to create new infrastructure and to contract with private providers. Among the key issues that the government will have to consequently address is the degree of autonomy allowed in the functioning of public providers. In addition, it will need to ensure fair competition between public and private providers, given that the latter will likely be large providers of services under any program to expand coverage. However, considerable challenges remain, particularly in ensuring access to health services in rural areas. There is a need for better integration of traditional (unqualified) providers who are often the only source of health care for populations in remote areas, as well as implementation of incentive schemes for trained medical personnel to stay in those areas.

The authors conclude the paper with a simple assessment of the resource implications of a policy intended to achieve universal coverage given the available data. India faces serious resource challenges in financing universal coverage, even neglecting long-term trends in demographics, income growth, and technological change that are driving health spending around the world. According to the authors' assessment based on adjustments to estimates presented in the literature, something of the order of 4.5 percent of GDP or higher is required to finance health services for the entire population, of which 1.1 percent will need to come from contributions by households. These estimates place the government contribution at a level higher than currently envisaged, including the projections of the HLEG report. This poses a serious revenue challenge to the government. Achieving a reasonable degree of access to health services for all of India's population is not a trivial task and there are many challenges that will need to be overcome to come closer to "universal health coverage."

The effectiveness of safety nets for the poor is increasingly coming under scrutiny in India, both due to the increasing fiscal pressure and concerns about poor targeting. The paper by Jha and Ramaswami explores problems with in-kind transfer programs by estimating the extent to which they reach the poor. The paper defines a percolation index to measure the effectiveness of food subsidies to the poor in India and the Philippines, which operate similar food-based, transfer schemes. The two countries spend significant resources on these. The cost of India's Targeted Public Distribution System (TPDS) amounts to around 0.65 percent of GDP. The Philippines' program, which is administered by the National Food Authority (NFA), amounts to about 0.3 percent of GDP.

The NFA program subsidizes rice, sourced mainly from monopoly imports, while the TPDS subsidizes rice and wheat, procured primarily from domestic farmers. The two programs have multiple mandates to balance producer and consumer interests. These include access to food by the poor, price stabilization through buffer stocks, and supporting farmers through prices that are often higher than prevailing market prices.

In both countries, government agencies source, store, transport, and distribute the grain to designated retailers who receive a fixed margin on the sale. The NFA program is not targeted, whereas the Indian program is. However, despite its universal nature, only about 16 percent of the population accesses the Philippines NFA program, partly because of self-targeting due to the supply of inferior quality rice and the unavailability of rice in some parts of the country. In India, all households are entitled to a monthly quota of 35 kg of rice or wheat. The magnitude of the price subsidy is the lowest for households above the official poverty line, a little higher for those below the poverty line, and the highest for the poorest.

While the literature recognizes that the gains to the poor depend on the efficacy of both targeting and of program delivery, most studies have only evaluated targeting performance. Their general finding is that most transfer programs are costly because of many nontargeted beneficiaries and that improving targeting involves both administrative costs and negative economic incentives. The common prescription for cost-effectiveness is to improve targeting by reducing inclusion errors and keeping the magnitude of the subsidy in check. The impact of inefficient program delivery on the poor has not received much attention, and when it has, the focus has been mainly on administrative costs, largely ignoring the cost due to corruption or theft. Jha and Ramaswami attempt to fill this gap by quantifying the extent to which food subsidy expenditures percolate to the poor based on both targeting leakages as well as nontargeting waste due to deficiencies in public program delivery that result in excess costs and fraud.

The authors ask the question: What is the impact of a marginal expansion of a food subsidy program on the poor? The percolation index is based on the breakdown of total expenditures on food subsidies into four components: income transfers to the poor; income transfers to the nonpoor (the leakage due to poor targeting); and two nontargeting leakages, the illegal diversion of subsidized supplies (pilferage), and the excess cost of the public distribution system over the cost of private sector distribution (the inefficiency of public distribution). The authors define the percolation index as the product of the share of the poor in the total cost of the subsidy (the depth of percolation, i.e., the extent to which expenditures reach the poor) and the participation rate of the poor (the breadth of percolation, i.e., coverage among the poor). The index varies between zero and one. If either of the two terms in the product is zero, then the index is zero. It takes a maximum value of one when all poor participate and when they receive the entire subsidy. When the participation rate is one, the percolation index is equivalent to the poor's share of the overall subsidy, the traditional measure of the incidence of targeting.

The authors show that neither the Indian nor the Philippines program scores well on the percolation index: Participation rates are low and households, whether poor or not, do not receive most of the expenditures of the food subsidy.

A principal finding of the paper is that the benefits from improving program delivery by reducing waste are much larger than the gains from lower inclusion errors through better targeting. Even though opportunities exist in both India and the Philippines for reducing targeting errors, this is not only contentious politically but is also subject to caveats in the literature. The literature offers examples where targeting is costly both administratively as well as in economic terms because of incentive effects. Some have argued that programs that are tightly targeted toward the poor (i.e., have low inclusion errors) do not receive political support from the nonpoor and are thus doomed to failure. In addition, there are the severe practical difficulties of targeting in poor governance environments. The authors suggest that problem with targeting in India is not so much that grossly ineligible households have been counted in but that many deserving households have been left out. The dilemma is that while a move toward enlarging the number of eligible households would increase participation rates and hence percolation, it would also increase inclusion errors and perhaps decrease the share of the poor in food subsidy expenditures.

On the other hand, it is more straightforward to recommend policies that reduce waste and improve efficiency. Even without improved targeting, the authors argue that the impact of food subsidies on the poor can be increased by raising their participation rate or by raising the fraction of the subsidy going to the poor, or a combination of both. Reducing the massive waste in the food subsidy programs of both countries (which the authors estimate at 65 percent and higher) provides potentially large cost savings that can be used to increase coverage substantially without a corresponding increase in public spending. The authors end with a discussion of the political economy of reforms in food subsidies and recommend a policy framework that allows for experimentation, learning, and adjustment even if such a framework runs against the bureaucratic impulse to use uniform formulaic mechanisms.

After experiencing low rates of inflation for several years, India has recently seen a return to near double-digit rates of price increase. Because this shift to higher rates of inflation has also been accompanied by relatively large increases in food prices, it has been a politically charged issue. In their paper, Prachi Mishra and Devesh Roy use the disaggregated monthly price data that underlie the Wholesale Price Index (WPI) to provide a detailed account of inflation over the last two decades, with a focus on food prices.

The authors begin by discussing the different indices that are officially published and outline how food price inflation may be measured. They discuss the trends in overall inflation and establish the commonly overlooked fact that low inflation (defined as 5 percent or less) has been a rare occurrence in the Indian economy in the last two decades. They argue that the long-term trend in the inflation rate exhibits a U-shaped pattern with a structural break in the year 2000 and an inflection point in 2002. Next, they show that the long-term trend for food inflation has also followed a U-shaped path, with a clear reversal of the declining trend starting in the early 2000s. Further, there is evidence for a moderate correlation between international and domestic food price inflation, albeit with significant variation across commodities based on their tradability. Moreover, the correlation is weaker when world prices are high than when they are low.

Mishra and Roy distinguish between food and nonfood inflation and find that food price inflation is typically higher, with the long-term trend for food price inflation consistently staying above that for nonfood inflation. They also find a significant pass-through from food to nonfood prices. A one unit shock to the annual rate of food inflation, which translates into a 0.2 percentage point increase in the long-run, is associated with at least a 0.05 percentage point long-run increase in the nonfood inflation rate. They argue that their analysis establishes the importance of food prices as a major driver of overall inflation in India.

The authors go on to disaggregate food prices into individual commodity prices and explicitly quantify the contribution of specific commodities to food price inflation. Although there is some variation over time, the top contributors typically turn out to be milk and fish in the category of animal-source foods, and the plant staples of sugar, rice and wheat, onions, potatoes, and to a lesser extent, edible oils.

Finally, Mishra and Roy conduct case studies of some items among the top contributors. The experiences of each of these commodities tell a distinct story about the factors explaining the inflation patterns and the policy responses. The following elements stand out: persistent demand-supply gap in milk owing to shifting demand by consumers; post Green Revolution fatigue in cereals; multipronged government interventions in sugar; high import penetration in edible oils; and inadequate infrastructure for smoothing prices across space and over time for fruits and vegetables.

The authors draw several policy implications. First, combining the macroeconomic and microeconomic analyses, a case can be made that in the context of a developing country like India, both the central bank and the other branches of government can play a meaningful role in tackling inflation. Given the high weight of food in overall inflation and the importance of supply factors or more structural and less interest-sensitive demand factors, the hope is that the RBI can control overall inflation primarily by curbing demand for more interest-sensitive nonfood items. However, even if the monetary transmission mechanisms operate efficiently, controlling nonfood inflation may only have a limited effect on overall inflation—particularly if the central bank envisages an inflation rate target that incorporates CPI-based indices, which have a relatively small weight on nonfood (31–54 percent).

The authors point to several mechanisms through which monetary policy can influence inflation. First, tighter monetary policy can appreciate the exchange rate and make the imported goods cheaper in local currency. Second, tighter monetary policy can reduce aggregate demand and hence indirectly reduce the demand for food. Finally, higher interest rates could raise the marginal cost of production, and therefore worsen the inflationary pressures (the so-called "working capital" channel). The last of these channels works against the first two and the relative strength of these mechanisms is an empirical question.

Mishra and Roy also note that their work has important implications for the choice of which price index the central bank should target for its measure of target inflation—headline (aggregate) or core (nonfood). The large weight of food in the consumption basket and the finding that food price inflation has been consistently above nonfood—both suggest a focus on the headline rate rather than various core measures. This is supported by further evidence of persistence in food price inflation, and the existence of some pass-through to nonfood. Furthermore, they argue that the CPI is the more appropriate measure of the headline inflation rate because it includes the items that are most relevant to consumers. Thus, a strong case for developing a more comprehensive CPI can be made.

Motivated by a balance of payments crisis in 1991, India has been engaged in a two-decade-long process of capital account liberalization aimed at expanding access to international capital markets. At the same time, it has sought to alter the composition of the capital inflows away from debt toward a greater emphasis on equity-based finance, promoting risk-sharing by foreign investors. The central bank's strategy has been to open financial markets cautiously while ensuring risk management systems and competencies are in place to strengthen the system's resilience to foreign shocks. Today, the equity market is relatively open to foreign investors, but the bond market remains quite closed. The primary objective of the paper by Renu Kohli and Agnes Belaisch is to evaluate the extent to which India's approach to regulation has insulated Indian markets from global shocks.

The paper begins with an extensive documentation of the complex regulatory regime governing foreign financial transactions. It highlights the degree to which the external borrowing of domestic firms is limited in amount, maturity, and cost. Foreign investment in domestic bonds is also restricted and requires registration as a foreign institutional investor (FII). In contrast, the rules governing foreign direct investment are quite liberal and foreign investors can acquire ownership stakes as large as 100 percent in most sectors. The rules for investment of FIIs in portfolio investments have also been relaxed, but remain subject to quantitative limits.

The authors estimate uncovered interest parity relationships and find that capital controls continue to provide a certain degree of autonomy to the central bank when setting policy rates. Vector Autoregressive (VAR) models are constructed to derive impulse response functions that trace out the transmission of a shock to foreign capital inflows throughout the different segments of India's financial sector, looking in particular at whether it spills over to the closed portions of the system. The paper exploits the differences in relative openness of stock, foreign exchange, debt and money markets within the structural VAR framework to answer the question of whether the capital account restrictions, especially upon debt flows, insulate the system from financial shocks. Uncovered interest parity (UIP) tests are used to evaluate the influence of the capital controls.

Through a three-pronged investigation of the relationship, Kohli and Belaisch find that UIP is routinely rejected by the data. First, the UIP regressions show that the domestic-foreign interest differentials are significantly different from zero for interest rates over a range of time horizons, suggesting imperfect integration. Second, the dynamics of UIP over time indicate that deviations of the uncovered interest differentials from their mean are permanent. And third, the speed at which the empirical interest differential converges to the nonzero "equilibrium" level suggests that capital controls play a significant role in slowing the speed of adjustment. The results imply that capital mobility, the equilibrating mechanism to restore the parity condition, is insufficient to lead to convergence. The comparison of their results for the relatively tranquil years of 2000–05 with the volatile period of 2006–09, suggests that the capital controls lost some of their effectiveness in the second period.

Impulse responses derived from the structural VAR model offer further evidence that capital restrictions on foreign participation in the Indian bond market insulate the bond market, but not fully. Arbitrage behaviors trigger a domestic channel of propagation between domestic financial markets. There is also evidence that central bank intervention temporarily dampens the exchange rate response to shocks to foreign equity flows. Partial sterilization leaves room for a response of interest rates, but with a lag. Inflation ends up higher than otherwise, but they find no significant impact on real variables. In sum, incomplete financial integration, resulting from existing capital controls in the bond market, provides a degree of short-term monetary autonomy to the central bank. Real economic activity also appears to be relatively insulated from financial shocks. From this perspective, the authors conclude that the regulations have promoted the system's resilience, and increased the autonomy of domestic monetary policy.

VIRAL ACHARYA
New York University Stern School of Business

Is State Ownership in the Indian Banking Sector Desirable?*

ABSTRACT A large part of the Indian banking system is still state-owned, at the same time when there are several private banks as well as nonbank financial institutions. This article provides evidence that the state ownership—by implicitly conferring stronger guarantees on state-owned banks—distorts the level playing field between various banking sector players. It then relates this lack of level playing field to the role played by government-sponsored enterprises (GSEs)—Fannie Mae and Freddie Mac—in the recent housing boom and bust in the United States. It therefore proposes a graceful exit of the state from the Indian banking sector.

Keywords: *Bailout, Government Guarantees, Government-sponsored Enterprises, Level Playing Field, State-owned Banks*

JEL Classification: *G01, G2, D6, K23*

I

The global crisis which began in the fall of 2007, and progressively worsened in 2008, affected the Indian financial sector beginning only 2008. While Indian financial firms have been fairly resilient compared to their global counterparts, Indian private sector firms faced greater losses compared to public sector firms during 2008–09. This was in spite of private sector firms having lower exposure to the crisis based on precrisis market indicators. By relocating their deposits, investors seemed to reward

* This article is based partly on the NYU–Stern Working Paper "State Ownership and Systemic Risk: Evidence from the Indian Financial Sector during 2007–09" (joint with Nirupama Kulkarni) and on the book *Guaranteed to Fail: Fannie Mae, Freddie Mac and the Debacle of Mortgage Finance* (coauthored with Stijn van Nieuwerburgh, Matthew Richardson, and Larry White; HarperCollins India, June 2011). I am grateful to Ajay Shah for providing the BSE and National Stock Exchange (NSE) stock market data.

public sector firms while penalizing private sector firms with similar risk. This should not be interpreted as greater resilience of state-owned banks vis-à-vis the private financial sector. It was access to implicit and explicit government backing rather than ownership by the state that helped public sector banks (PSBs) perform better.

II

In 2008, the global financial crisis hit India with the Indian stock market losing more than 60 percent of its peak valuation. Figure 1 shows that the stock market index—S&P CNX NIFTY index—declined sharply starting January 2008. Index prices fell from a peak of 6,288 in January 2008 to 2,524 in October 2008, representing a decline of nearly 60 percent. Another market index—the Bombay Stock Exchange (BSE) index—similarly fell nearly 59 percent from 20,873 in January 2008 to 8,510 in October 2008. Starting 2008, foreign institutional investors (FIIs) facing a liquidity squeeze from abroad, started pulling out capital from India resulting in a sharp decline in the stock market. In 2008–09, FIIs withdrew nearly ₹43,337 crores (approx $9–10 billion).

Experiencing a dearth of capital from overseas markets, Indian banks and corporations had to turn to domestic markets for their funding requirements. At the same time, Indian banks and financial institutions facing uncertain market conditions started cutting back on credit, resulting in a liquidity crisis in 2008. Corporations, especially ones relying on foreign funding, feared further worsening of global market conditions and withdrew from money market mutual funds (MMFs). The MMFs, which were heavily invested in nonbanking financial companies, were forced to liquidate their positions. It is estimated that MMFs withdrew nearly ₹22,355 crores in 2008–09. As a result of the capital outflows, the rupee also came under pressure. There was further liquidity tightening as the Reserve Bank of India (RBI) intervened in the Forex market to manage rupee volatility. All these events resulted in a money market and credit squeeze which eventually spilled over into the real economy (Subbarao, 2009). The global slowdown also resulted in a slump in demand for exports. This impact was felt economy-wide and the Indian GDP fell from 9 percent in 2007 to nearly 6.1 percent in 2008. Eventually, the Government of India, fearing an even rapid deterioration of the economy, announced wide-ranging stimulus packages in 2009 that appeared to restore the economy back to its pre-2008 growth.

FIGURE 1. Stock Index Performance

The figure shows stock index performance for the period from January 2007 to February 2009. Two indices, S&P NIFTY and BSE SENSEX, are represented. The S&P CNX NIFTY (or NIFTY; base level of 1000 defined as of November 1995) is a free-float market capitalization index on the National Stock Exchange and consists of 50 companies. Bombay Stock Exchange Sensitive Index (BSE Sensex or Sensex) is a value-weighted index composed of 30 stocks with a base level of 100 in 1978–79.

An important observation during 2008 was the apparent weakness of private financial firms against the (relative) growing strength of public sector or state-owned banks. Historically, Indian banks had been wholly owned by the government. Though in the 1990s, after economic liberalization, the government reduced its stake and allowed private banks and foreign players to enter the market, the Indian financial system retains a substantive public sector ownership. In fact, PSBs dominate the Indian banking sector and, as of March 2009, they accounted for nearly 71.9 percent of aggregate assets. This mixed model of public and private ownership, popular in emerging markets and also referred to as the Asian model, has been credited with the relative strength of the Indian financial sector compared to its global counterparts.

Consider three striking pieces of evidence illustrating this relative strength.

First, and as Figure 2 shows, the substantial gains made by the private financial firms in the period leading up to January of 2008 were almost entirely wiped out by February of 2009.

Second, market reaction to public versus private sector banks can also be gauged from the widening of credit default swap (CDS) spreads for two illustrative firms, namely, State Bank of India (SBI, a PSB) and ICICI Bank (a private sector bank) during the crisis of 2008. A CDS spread represents the cost of purchasing insurance against the default of an underlying entity (such as SBI or ICICI). From Figure 3, we see that the cost of purchasing one-year protection on SBI and ICICI banks were within the same range in 2007, suggesting that investors viewed both firms as being equally risky. Beginning 2008, however, the difference between the spreads started widening in SBI's favor, indicating that investors possibly viewed the public sector financial firm to be healthier compared to the private sector firm.

Third, Figure 4 shows that while the banking sector as a whole experienced a slowdown in deposit growth, private sector banks were affected to a larger extent. According to RBI estimates, PSB deposits grew by 24.1 percent in fiscal year 2009 (March 2008–March 2009) compared to 22.9 percent a year earlier. In comparison, private sector deposit growth slowed from 19.9 percent to a mere 8 percent for the same period. Credit growth showed similar trends: PSB credit grew by 20.4 percent (compared to 22.5 percent in 2008), while for private sector banks, it grew by only 10.9 percent (compared to 19.9 percent in 2008).

This apparent strength of PSBs has in fact strengthened the resolve to persist with them. Until recently, there had been a consistent trend toward

FIGURE 2. Stock Performance for Public and Private Sector Firms

The figure shows the equally weighted average returns for public and private sector financial firms for the period from January 2007 to February 2009. The indexed value weighted returns for the private (public) sector represents the returns weighted by the market capitalization of the private (public) sector firms used in the analysis. Both series use a base value of 100 as of January 2, 2007.

FIGURE 3. CDS Spreads for ICICI and SBI

The graphs below show the 1-year and 5-year CDS spreads for ICICI Bank (private sector bank with the largest MES) and State Bank of India (PSB with the largest MES). Market return is based on the S&P CNX NIFTY for the precrisis period from January 2007 to December 2007.

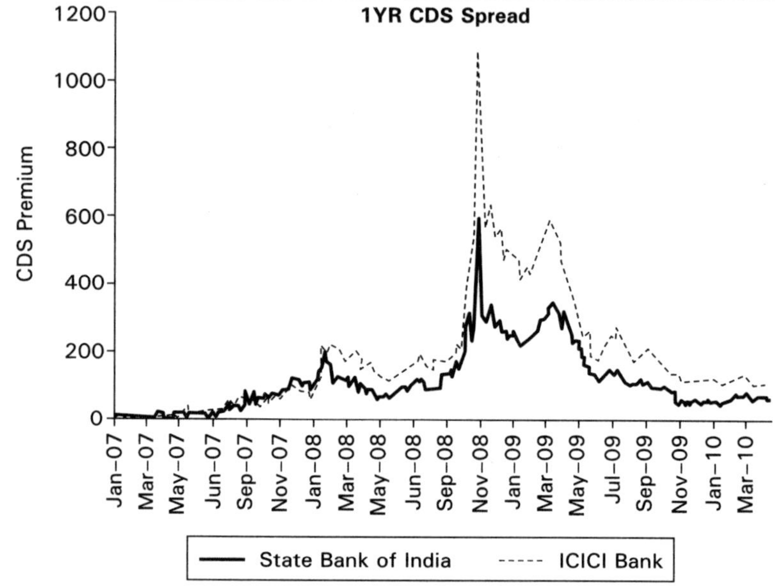

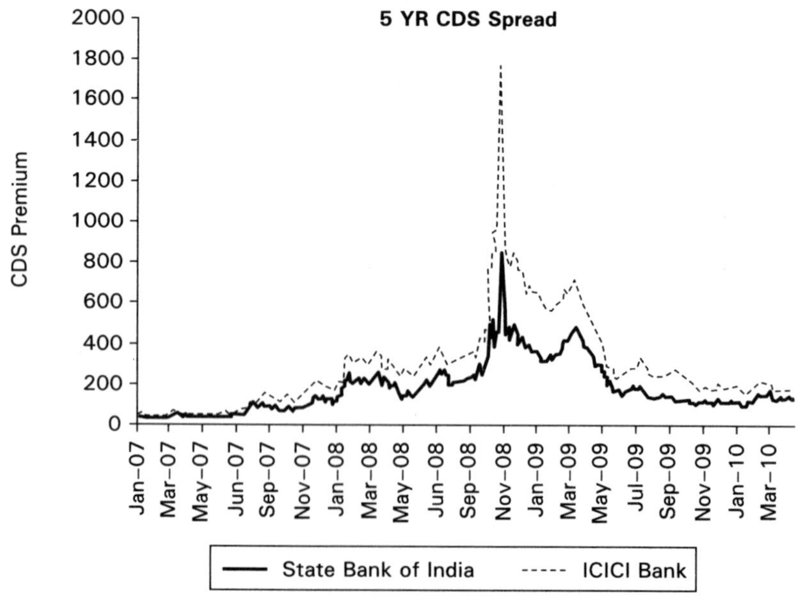

Source: Datastream.

FIGURE 4. Deposit and Credit Growth

The graphs below show the group-wise growth in deposits and credit in banks. Growth rates are year-on-year (y-o-y) as of March 28, 2008 and March 27, 2009.

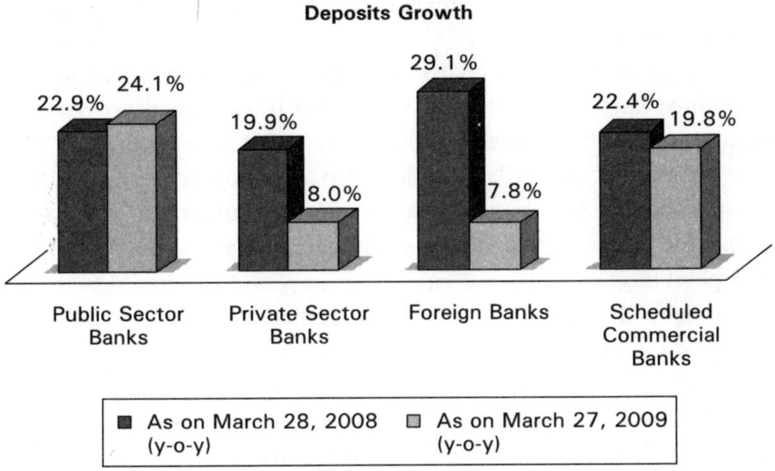

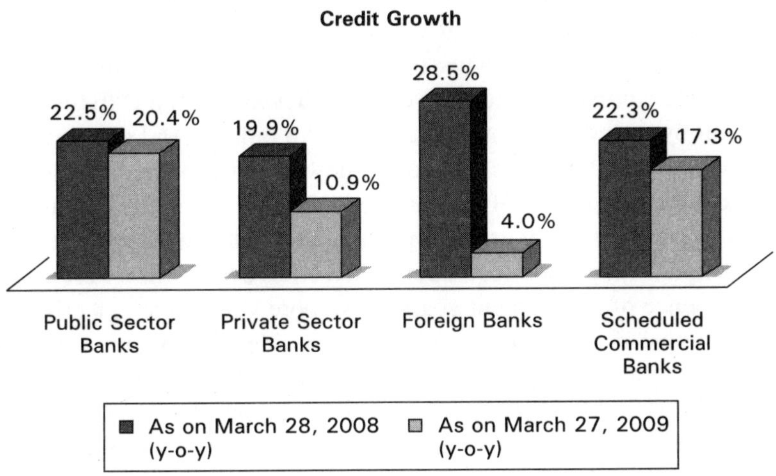

Source: RBI, 2009–10.

privatization of the Indian banking system. However, the recent underperformance of private sector banks has raised some doubts regarding this approach.[1] Such sentiments have important policy implications and could

1. The ruling party leader, Sonia Gandhi, claimed that "public sector financial institutions have given our economy the stability and resilience we are now witnessing in the face of

alter the timeline and extent of privatization initially envisioned by the government. All recent evidence seems to suggest that government ownership in PSBs will gradually decline, but only after the ongoing crisis has subsided, and that it is unlikely the state-owned banks will be fully privatized as was previously envisioned.

There is good reason to guard against reaching such policy decisions. As explained later in this essay, it seems that it was access to government guarantees rather than ownership by the state that appears to have ensured greater stability of the PSBs than their private sector counterparts. In fact, PSBs appeared *more* risky in the sense of being vulnerable to a crisis, not less, than the private financial firms, and yet grew substantially in terms of their deposit base.

The conjecture is that the relative underperformance of private sector banks during the crisis, in spite of their superior precrisis risk–return profile, is attributable to the implicit and explicit sovereign backing of PSBs. The Indian Bank Nationalisation Act provides an explicit guarantee that all obligations of PSBs will be fulfilled by the Indian government in the event of a failure. It is conceivable that, as a result of this guarantee, private sector banks experienced a loss of confidence and capital gravitated to PSBs during the crisis of 2008–09—even when their exposures to an economy-wide crisis were ex ante similar—because investors believed that the PSBs would be bailed out by the government in the event of a failure. And that, given this expectation, capital (primarily deposits) flew from the riskier private sector banks to the more stable PSBs, resulting in a decline in equity valuations of the private sector financial firms during the crisis. If this were true, then PSBs would outperform private sector banks in a crisis even if they were a priori, i.e., in absence of government guarantees, more vulnerable to a crisis.

To examine if this conjecture is true, consider the ex ante (pre-2008) measure of vulnerability to a crisis (*systemic risk*) of public and private financial firms and how that relates to their ex post (2008–09) or realized performance—for instance, their deposit growth. We use the marginal expected shortfall (MES) measure to calculate the systemic risk of financial institutions in the Indian financial sector.[2] The MES measure essentially

the economic slowdown." Then finance minister, P. Chidambaram, echoed these sentiments when he claimed that India's public sector banks were strong pillars in the world's banking industry (Chandrashekhar, 2008).

2. The strength of the measure lies in its ability to predict which firms are likely to be most negatively affected during a financial crisis, as demonstrated by Acharya et al. (2010) in their analysis of the systemic risk of large US financial institutions around the financial crisis of 2007–09.

captures the tail dependence of the stock return of a financial firm on the market as a whole. It estimates, in a given past period (say one year preceding a crisis), for the worst 5 percent days of the market or the financial sector index, the negative of the average market return of a given financial firm. The greater the MES, the more vulnerable to a crisis is the firm. The question then is whether riskier PSBs as measured by ex ante MES fared better or worse than private sector banks with similar risk.

There are some interesting insights to be gathered from such analysis.

First, average MES value measured during January to December 2007 was higher for PSBs (4.34 percent) compared to private sector banks (3.58 percent). That is, the PSBs had, on average, negative 4.34 percent returns on the days the market return (S&P CNX NIFTY) was below its fifth percentile for the precrisis period from January 2007 to December 2007. India Infoline (6.99 percent), IFCI (6.80 percent), and Indiabulls financial services (6.44 percent) had the highest MES among the private sector financial firms. In the public sector, IDBI Bank (6.67 percent), Union Bank of India (5.41 percent), and Dena Bank (5.23 percent) had the highest MES. Focusing next on losses in terms of rupees, in the PSBs, SBI had the highest market capitalization loss of ₹30 crores, whereas for private sector banks, ICICI had the highest loss of ₹37 crores.

Second, PSBs performed *better* than private banks during the crisis in spite of having *higher* systemic risk. While private banks with high MES prior to the crisis (such as ICICI Bank) suffered heavily during the crisis, equally systemic state-owned banks (such as the SBI) gained in a relative sense. For example, both ICICI and SBI had an MES of 5 percent. However, during the crisis period from January 2008 to February 2009, ICICI stock fell by 73 percent, whereas SBI stock fell by a significantly lower 54 percent.

Third, while private sector banks with higher vulnerability to a crisis experienced deposit contractions, the reverse was true for PSBs. Using the RBI deposit flow data, Figure 5 shows the quarterly change in deposits for public and private sector banks. We see that when the crisis initially hit India in 2008, both public and private sector had similar deposit growth rates. In Q1 2008, deposits for both sectors grew by 10 percent. However, as the crisis worsened, the disparity between public and private sectors is evident. PSB deposits grew by 1.7 percent (Q2), 5.5 percent (Q3), and 5.2 percent (Q4) compared to a much lower growth of 0.0 percent, 1 percent, and –0.3 percent for private sector banks.[3] This suggests that investors treated public

3. Toward the end of the crisis, both sectors posted relatively higher growth rates of 12 percent for the public sector and 8.2 percent for the private sector, a fact that can be explained by government stimulus package discussed later.

FIGURE 5. Q-o-Q Deposit Amounts and Growth for Public and Private Banks

The graph below shows the quarter-over-quarter (q-o-q) changes in deposit amounts for public and private sector banks. Deposit amount data is in INR crores for the period Q1 2007 to Q1 2009.

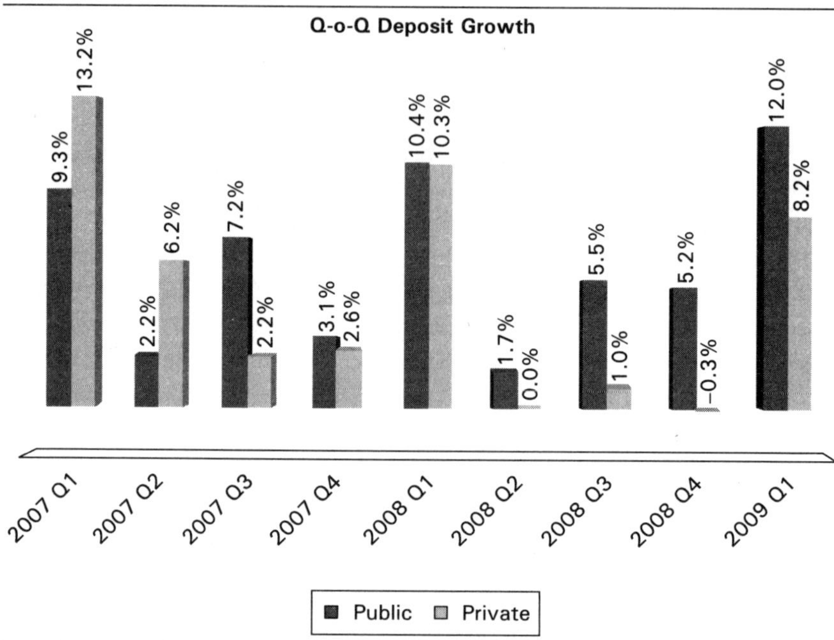

Source: RBI.

and private firms differently during the crisis. In particular, there was a shift in deposits from private sector firms to the public sector.[4]

And fourth, perhaps most strikingly, Figure 6 confirms that PSBs benefited from deposit growth even when they had greater vulnerability to a crisis.

Graph A shows that vulnerability to a crisis (MES) does a good job of explaining the growth in deposits for private sector firms. As we would expect intuitively, private sector banks with high exposure to a crisis experienced greater deposit contraction during the crisis. A few cases illustrate this

4. Anecdotal evidence is consistent with this "flight-to-quality": Infosys transferred nearly ₹10 billion of deposits from ICICI to SBI just after Lehman's collapse in the third quarter of 2008 (*Economic Times*, 2009): *"SBI was the biggest gainer in the 'flight to safety' phenomenon observed after the collapse of Lehman. With rumours about banks being in trouble, investors shifted funds away from foreign and private banks into government banks ... In the post-crisis quarter, even large corporates like Infosys moved their deposits to SBI. Infosys has disclosed that it transferred deposits of nearly ₹1,000 crore from ICICI Bank to State Bank of India last year."*

FIGURE 6. Deposit Growth versus MES for Private and Public Banks

The graphs below show the scatter plot of the MES computed for the period January 1, 2007 to December 31, 2007 versus the deposit growth for PSBs from March 31, 2008 to March 31, 2009. Market return is based on the S&P CNX NIFTY for the precrisis period from January 2007 to December 2007. Deposit growth for the crisis period is measured from March 31, 2008 to March 31, 2009. The 39 firms for which both MES data and RBI deposit growth estimates are available were used in this analysis.

Graph A: Private Banks

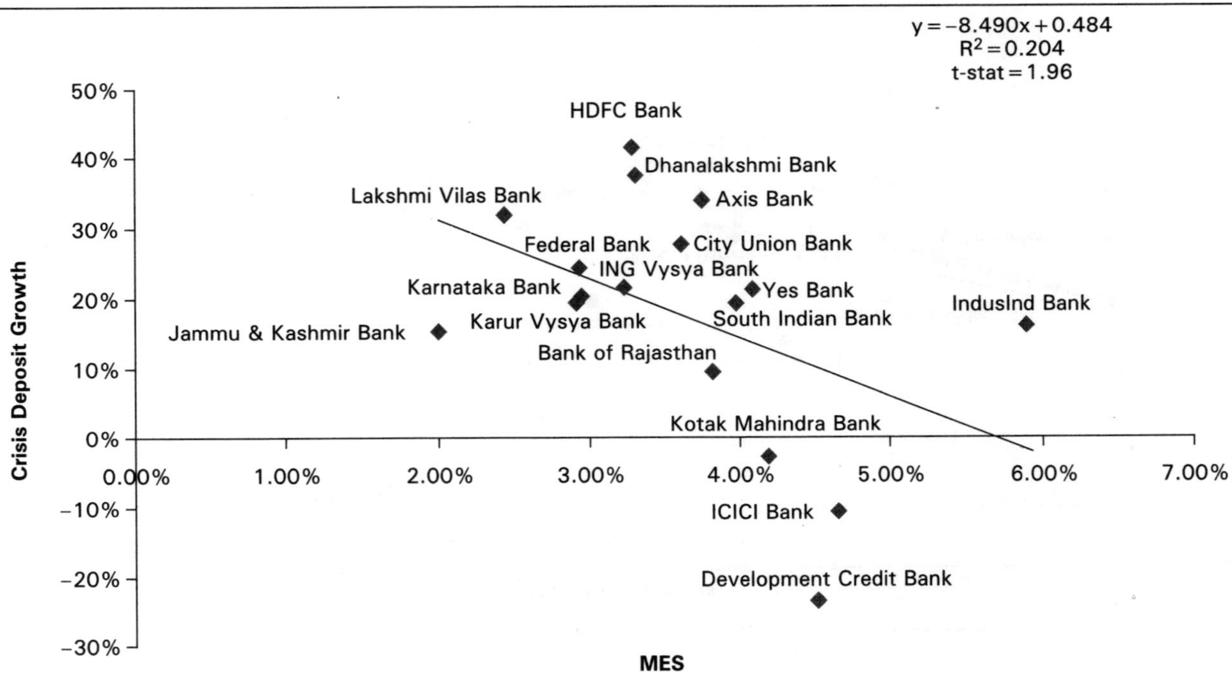

Graph B: Public Banks

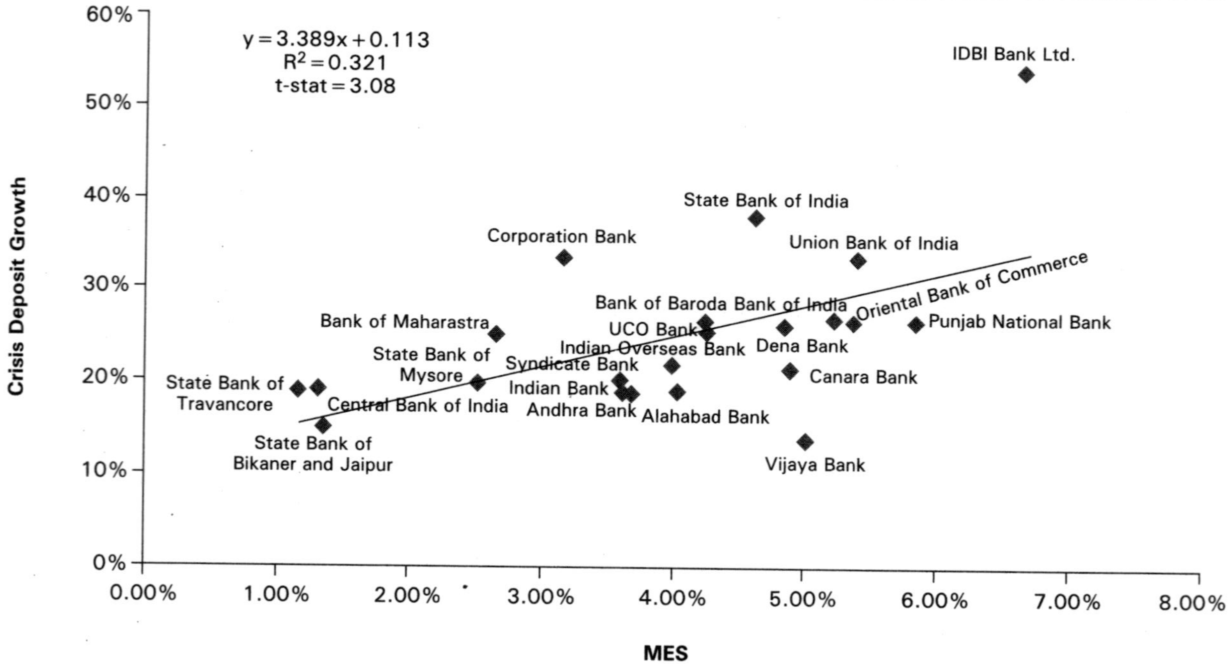

Source: Acharya and Kulkarni (2012).

point well. Indusland Bank with a high MES of 5.90 percent had a deposit growth of 16 percent in the crisis period. Compared to this, Axis Bank, with a relatively lower systemic risk exposure (MES of 3.75 percent), had a higher growth rate of 34 percent.

Graph B shows that, for PSBs, there is a somewhat counter-intuitive finding: greater vulnerability to a crisis in fact led to greater deposit expansion! Looking at specific examples, we see that deposits for SBI (MES of 4.63 percent) grew by 38 percent whereas, in contrast, deposits for Andhra Bank (with a lower MES of 3.61 percent) grew by only 20 percent.

Is it possible that the government backing of public sector firms distorted investor and market behavior during the crisis, rewarding public sector firms with greater vulnerability to the crisis, given the greater likelihood that more vulnerable banks will be bailed out in the event of a failure during the crisis? Evidence does suggest that weaker PSBs received capital injections, in whose anticipation depositors and stock market investors rewarded riskier PSBs while penalizing private sector banks with similar risk.

When the Indian government announced a number of wide-ranging stimulus plans to jumpstart the banking system, PSBs were promised capital injections to help them maintain a CRAR (risk-adjusted capital ratio) of 12 percent. The government launched three fiscal stimulus packages during December 2008–February 2009. As part of the second stimulus package, the government recapitalized state-run banks and infused nearly ₹3,100 crores in 2008–09 as tier-I capital in a few PSBs. In order to fulfill the funding gap, the government requested financing of ₹1,700 crores ($3.4 billion) from the World Bank in December 2008. Importantly, the timing and size of the capital injections was left up to the discretion of the government. Capital injections were to be determined based on PSBs' ability to access equity markets, capital requirements for growth, and existing capital resources (World Bank, 2009). An additional infusion of ₹16,500 crores was projected for the year 2010–11 to help PSBs maintain the minimum 8 percent tier-I capital to risk weighted asset ratio (Government of India, press release, 2010).

Investigating who received support tells the familiar story of worst performers being propped up. In February 2009, the government announced a capital injection in three PSBs: UCO Bank (₹450 crores), Central Bank of India (₹700 crores), and Vijaya Bank (₹500 crores). For the 2008–09 period the government injected a total of ₹250 crores into United Bank of India. The government also announced capital infusion of ₹6,121 crores in five more PSBs: IDBI Bank (₹3,119 crores), Central Bank (₹2,016 crores), Bank of Maharashtra (₹590 crores), UCO Bank (₹375 crores), and Union Bank (₹111 crores).

As of March 2009, all these banks (except Union Bank) had tier-I capital less than 8 percent: Bank of Maharashtra (6.1 percent), Central Bank of India (7.0 percent), UCO Bank (6.5 percent), Union Bank of India (8.2 percent), Vijaya Bank (7,7 percent), and IDBI Bank (6.8 percent). Based on the MES measure of vulnerability to a crisis, these were also among the riskiest banks in our analysis. For example, IDBI had an MES of 6.67 percent; Union Bank of India, 5.41 percent; and Vijaya Bank, 5.02 percent. UCO had a relatively lower MES of 4.26 percent. Indeed, IDBI with a high MES of 6.67 percent received the highest capital injection of ₹3,119 crores.

With such generous backing of the Indian government, PSBs came out in the retail sector with inexpensive housing, auto, and education loans. For example, they were the lead financers in the Tata Nano auto purchases. They were also able to offer housing loans at lower rates than those charged by other banks and mortgage companies, such as Housing Development Finance Corporation (HDFC). Private financial firms, in fact, have complained that SBI schemes do not draw in new customers, but are instead targeted at existing customers and are, thus, targeted more to undercut competitors rather than stimulate the economy (Wharton, 2009). All this suggests that the state banking sector may have grown during the crisis at the expense of private banks. Measures taken by the government may have helped bolster PSBs, but they have also made it difficult for private sector financial firms to compete with them.

III

That financial firms with access to explicit and implicit government guarantees for public sector firms fared better *during* the financial crisis has, in fact, been the theme worldwide: these firms survived the crisis or even expanded post-crisis while the ones without such access have failed or shrunk. A striking case in point has been the growth of the GSEs (Fannie Mae and Freddie Mac) and commercial banks in the United States—both sets of institutions with explicit government support and ready access to central bank emergency lending. These institutions expanded their holdings of mortgage-backed securities during the crisis while investment banks and hedge funds de-leveraged and sold these securities. Fannie Mae and Freddie Mac were hardly the better-performing institutions of this crisis based on their precrisis risk-taking behavior; they were, in fact, "guaranteed to fail" (Acharya et al., 2011).

So what are the similarities between the United States and India as far as government involvement in finance is concerned? Are there lingering risks for India of the form that were manifested for the United States through the risk-taking and eventual collapse of Fannie and Freddie? What preventive—or corrective—measures can India undertake in order to avoid a government-sponsored crisis? Or is India, the democracy with the greatest growth potential over the next few decades, destined to run into similar problems as United States, its democratic counterpart of the last century?

First and foremost, it is important to recognize that India is a tale of two parallel economies—not the white-money and the black-money economies, as immediately comes to mind—but of fast-paced urban India and sluggish rural India. Urban India provides most of the significant manufacturing and services growth to the economy; has affluent families even if coexisting with highly poor ones; and exhibits reasonable levels of access to bank savings, stock markets, mutual funds, insurance products, and the like. Rural India, in contrast, is farming-focused, is relatively poorer with current earning that is less than $2 per day, and has abysmal levels of financial inclusion. While agricultural wage labor does have access to informal savings schemes, that is a second-best solution (in the economic parlance), rather than the first-best. Given such a lack of financial inclusion in rural India, comprising over 75 percent of India's population by most estimates, there is both an economic and a political case for improving access to finance. How can the government achieve the transition of rural India in terms of growth and financial access to levels that are at least comparable to those of urban India, if not of the most developed economies?

Unfortunately, answers in India to questions of this sort have by and large been of brute-force interventions, rather than a nuanced enabling of the private sector in aiding such a transition. The experiment of Fabian Socialism sapped the private sector of its growth potential for close to four decades post independence, and continued to leave an indelible imprint even through the reforms of the 1990s. Political forces ensured that a series of successive governments, under one pretext or the other, preserved the government hegemony over a large number of business segments (oil, natural gas, power, aviation, among others) that have yet to be privatized or are, at best, poorly privatized. While this imprint in itself would call for a separate treatment, more to the point of this book have been the government efforts to improve financial inclusion.

This is where the Indian government-sponsored enterprises in the financial sector come into play. Currently, the Indian banking sector is partly state-owned. State-owned banks own over 70 percent of banking assets;

until the 1990s reforms, the state ownership of the banking sector was in fact 100 percent. Indeed, India is one of the nine countries in the world to have a predominantly state-owned banking sector (the others being Bhutan, Libya, Algeria, Belarus, Turkmenistan, Egypt, and Costa Rica). Among fast-growing emerging markets, its state presence in banking is paralleled only by China. The state-owned banks in India enjoy privileged branching, especially in rural areas, where the private sector, often eager to enter, is by and large restricted from entry, or at least not allowed to enter by default.

The origins of such desire for state control of rural finance are many. It is partly rooted in the socialist view of the economy that took immediate hold following independence, but it is largely a matter of political convenience. It is important to keep over 75 percent of India's population happy, at least in the short term, when they are about to cast important votes. Not surprisingly, a study by Cole (2009) found that agricultural lending in India has a cyclical pattern that is not coincident with the economic cycle of India, but is coincident with the political cycle instead.

There are distinct parallels between this foray of the Indian government into rural banking and the role of Fannie Mae and Freddie Mac in the United States.

First, affordable home ownership goals of the United States (applicable to low-income and thereby poor–credit quality households) relate to "priority sector lending norms" in India (focused on agriculture and small-scale industries, but also include housing).

Second, the GSEs in the United States underwrote over 50 percent of mortgage credit risk in the residential housing market. This corresponds to the large, close to 70 percent, share of state-owned banks in India.

Third, the forbearance that was exhibited toward the GSEs—in good times, through generous capital requirements (2–2.5 percent of capital against underwriting the credit risk of residential mortgage pools versus 4 percent for private sector) and in bad times, through government conservatorship—mirrors closely the reluctance of the Indian government to shut down any state-owned bank as discussed in section II (or for that matter, even pull down shutters on rural cooperatives that also aid the priority sector lending goals).

Fourth, the GSEs crowded out the private financial sector in prime (high credit quality) residential housing market and eventually also in affordable housing loans. So have the state-owned banks in India crowded out the private-sector counterparts. The annual incomes of bank customers in private and foreign banks in India are substantially greater than those of the customers of public sector and cooperative banks; and yet, like the quality of subprime loans of GSEs, the nonperforming assets from priority sector loans are greater for state-owned Indian banks than for private and foreign banks in India.

The Indian state has thus freely traded in credit extension to the poor in a manner that is similar to that of the United States government. Alas, it too has condemned its poor to be even poorer in the process. A bulk of the priority sector lending in India goes to farming loans. Since banks are effectively forced to lend at quantities that they would not lend had they been entirely private and free of state-enforced lending norms, the quality of the lending is poor. Put simply, lending happens down the quality curve up to a point that is not worthy of finance. In turn, even low-productivity farming and agricultural investment in rural India can be readily financed. Of course, loans in such investments go bad, but most of the losses are concentrated in state-owned and cooperative banks. The strong clutch of the government now becomes a crutch, readily extending forbearance where the invisible hand of the market would result in shutting down of banks. Since these banks never fail, they do not care about the quality of their loans. Since they do not exercise due diligence with respect to borrowers, the farmers do not invest in high-productivity and innovative agricultural methods. And, of course, they do not consider retooling to any other skill set besides farming. Thus, neither the banks nor the farmers are subject to the capitalist creative destruction that would seem essential to success in a country where the rural, farming-focused sector is so large and yet low on productivity.

One wonders in fact if the demographic dividends that India is supposed to reap in the next few decades, by virtue of an anticipated decline in its ratio of the dependent to the working-age population, will be more than offset by the negative dividends resulting from the poor quality of farming? In a country of over 1.21 billion people by the latest census, with a rapidly expanding middle class, investments in productive farming ought to find a center stage. Yet, India stumbles from year to year, facing acute food shortages that fluctuate with the vicissitudes of tropical monsoons and create a never-ending predicament for the RBI, India's central bank, in whether to fight food inflation through interest-rate rises or aid the economy's reaping of its demographic dividends through interest-rate cuts. As RBI's deputy governor, Subir Gokarn (2011), comments:

> Food inflation, covering both unprocessed and processed items, already at a relatively high 10.4 per cent in June 2009 (when the headline rate troughed) soared to 20.2 per cent in December 2009. It has come down since then, but has remained high at 6.4 per cent in February 2011 … All along, the objective was to not disrupt

> the growth recovery, while containing the spread of inflationary pressures from food and energy into a wider process. The Reserve Bank, therefore, chose a steady, calibrated response. Given the significance of supply side inflationary pressures throughout the period, stronger interest rate actions might have slowed growth far more than they would have slowed inflation.

Another concern arises by understanding how Fannie Mae and Freddie Mac eventually failed. There was a "race to the bottom" between the public sector and the private sector in the United States finance. In their important book on the American corporation, Berle and Means (1932) introduce the concept "race to the bottom": the idea that competition can lead to a reduction of standards. While Berle and Means were referring to regulatory standards and competition among the (then) 48 states, it is not difficult to see how the same arguments could be applied to government-sponsored enterprises and likewise financial institutions.

Figure 7 graphs the tremendous growth in the mortgage market in the United States, and the fraction of residential mortgage originations each year that were securitized by the GSEs or private-label firms, as well as the amount not securitized. As can be seen from Figure 7, the mortgage market

FIGURE 7. Growth in Mortgage Market, Securitization and Percentage Share of Market in the United States

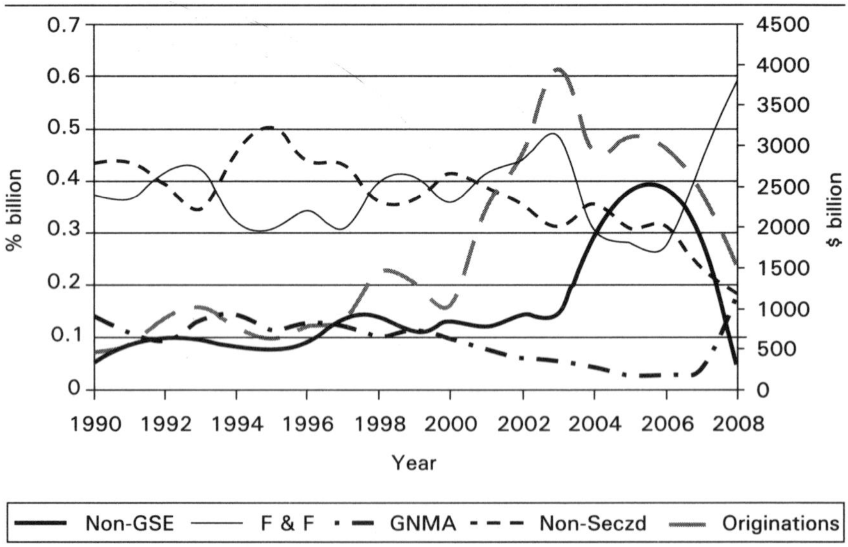

Source: Acharya et al., 2011.

Note: GSE = government-sponsored enterprise.

increased dramatically in size, especially in the latter 2003 period, with the emergence of the riskier subprime and Alt-A mortgage lending. Figure 7 shows that mortgage securitization generally increased every year from 1995 onwards, albeit for different reasons in different periods. In the period up to 2003, the GSEs dominated the market, but, then post 2003, non-GSE MBS more than tripled from 12 percent to 38 percent of the origination market. By 2006 and 2007, the Figure illustrates well the competitive struggle as the GSEs recovered their market share.

Was the growth in private-label MBS the culmination of the dream of the 1982 "Commission on Housing" in the United States coming to fruition, albeit two decades later? Or was it the emergence of new government-sponsored enterprises in the form of too-big-to-fail financial institutions? Acharya et al. (2011) convincingly argue for the latter, describing the battle between the GSEs, whose accounting problems in 2002–03 slowed their balance sheet growth, and the too-big-to-fail large complex financial institution (LCFI), who first gained market share when the GSEs slowed but then eventually lost their share by 2006 again to the GSEs which had recapitalized their losses by then. Along the way of this race to the bottom in mortgage finance, underwriting quality deteriorated across board, setting up the stage for an unprecedented housing boom that crashed in 2007.

Paradoxically, the mother of all failures—those of Fannie and Freddie—ended up with these entities' owning the mother of all shares in the mortgage finance post crisis in the United States, so much so that over 90 percent of mortgages currently being issued in the United States are guaranteed by some government-sponsored enterprise. Something similar, albeit on a smaller scale, happened in India in 2008 as well, as the genie of government guarantees surreptitiously crept up, stabilizing the banking sector but eroding the private banks. The genie needs to be put back in the bottle now, but government guarantees are unfortunately being anointed as the pillar of modern Indian banking.

Let us rewind to 2008. The crisis hit India somewhat unexpectedly. In early 2008, there were talks of how "decoupled" emerging markets such as India and China had become from the financial and economic malaise of the Western economies. However, the interconnected nature of the global financial plumbing debunked any such myth rather soon. In the case of India, a number of large multinational firms had external bank and corporate bond borrowings in foreign currencies and markets. As the Western financial sector started unraveling in early September of 2008 and Lehman Brothers filed for bankruptcy on September 15, the market freezes abroad created a severe liquidity shortage for these Indian firms. Their immediate response was to ask for redemption of their savings from the Indian money market funds. As these funds faced liquidations of their portfolio holdings, they also withdrew from their lending to financial institutions, including banks. The precautionary desire of Indian corporations, stemming from the market freeze abroad, effectively began to translate into a freeze in the Indian interbank markets. While all this was happening, the stock market had already witnessed a 50 percent drop from its peak at the end of 2007, and some private sector Indian banks with foreign exposures were under watch by investors for a potential run. What ensued was in many ways magical in the short run but, as described empirically in section II, entrenched the stronghold of public sector banking in India even further.

Savers in the Indian economy, fearing a banking crisis and runs on the private sector banks, started moving their deposits to PSBs. As per RBI estimates (also shown in section II), PSB deposits grew by 24.1 percent during fiscal year 2009 (March 2008–March 2009) compared to 22.9 percent growth a year earlier. In comparison, private sector deposit growth slowed from 19.9 percent to a mere 8 percent for the same periods. In anecdotal evidence that is consistent with this "flight-to-government-guarantee," Infosys transferred nearly ₹10 billion of deposits from ICICI Bank (a private sector bank) to the SBI (the largest PSB) just after Lehman's collapse. And unsurprisingly, when the Indian government announced a number of wide-ranging stimulus plans to jumpstart the ailing banking system, several weak PSBs received a significant share of stimulus to restore their capital levels.

The flight of deposits in this episode appears to have stacked the odds against the private financial sector considerably since. They are, in effect, getting somewhat crowded out. The credit grew for PSBs by 20.4 percent in 2009 (compared to 22.5 percent in 2008), whereas for private sector banks, it grew by only 10.9 percent (compared to 19.9 percent in 2008). In the retail segment, PSBs came out with inexpensive housing, auto, and education loans, emerging as the lead financers in the Tata Nano auto purchases. The ruling party claimed that "public sector financial institutions have given our economy the stability and resilience we are now witnessing in the face of the economic slowdown." And while such stability and resilience has converted—perhaps just reinforced—Indian PSBs into the equivalent of GSE Godzillas, who now want larger share of all intermediation markets including insurance, the private sector King Kongs have recently complained—as they did in the United States about Fannie and Freddie's low-cost funding—that PSB schemes are primarily undercutting competitors, given the former's implicitly guaranteed funding base.

All evidence presented in section II suggests that the state-banking sector in India grew during the crisis at the expense of private banks. The government, however, has yet to recognize—or acknowledge—that the growth of state-owned banking sector since 2008 has been the effect of state guarantees, *not* state ownership. Banks all over the world were stabilized—whether public or private—once governments backstopped them. This emergency backstop, the emergence of the genie out of the bottle, need not, however, be coincident with the misfortunes associated with state ownership in good times. One, state ownership creates severe moral hazard of directing bank lending for politically expedient goals and of bailouts when such lending goes bad. Second, state ownership restricts the ability of state-owned banks from raising arm's length capital against state's stake, strangling their growth and keeping these banks—and certainly their private capital base—smaller than it need be.

Until recently, there had been a consistent trend toward the privatization of the banking sector in India. However, the recent underperformance by private sector banks has raised some doubts regarding this approach, and created a political consensus against the precrisis notion that the state-owned banks be fully privatized in due course. Many see this as another nail in the coffin of efficient privatization of sectors with significant presence of the Indian state. The government-sponsored enterprises can swiftly morph into hedge funds of the likes of Fannie and Freddie that are *guaranteed to fail*. This should strike a note of caution to the stance of Indian policymakers against drawing conclusions favoring a state presence in the financial sector.

Indeed, an Indian GSE has been before where Fannie and Freddie are now. The free trade by the Indian state historically extended beyond credit markets. In 1963, Indian state created the Unit Trust of India (UTI), which for more than two decades remained the sole vehicle for investment in the capital market by the Indian citizens. In mid 1980s, PSBs were allowed to open mutual funds. UTI rode the stock market rise of 1990s till the "Ketan Parekh scam" burst the market. UTI's flagship and largest scheme, US 64, had promised returns as high as 18 percent over a period going up to two decades. A full-fledged run on the Indian stock markets would have likely ensued had the government not come out with a rescue package. UTI Act was repealed, UTI was broken up, and UTI Mutual Fund still runs, but now like any other mutual fund, controlling less than 10 percent share of the market and operating at a distance from the government. UTI, like Fannie and Freddie, was just that—guaranteed to fail—a government-sponsored hedge fund that brought about a government-sponsored crisis.

And while it has not led to a significant crisis so far, the Life Insurance Corporation (LIC) of India, the largest state-owned life insurance company, is again also the country's largest investor. The penetration of its insurance business in rural India, which might seem to be the primary goal of state owning such an enterprise, is far less impressive, than its penetration as an investor, funding close to 25 percent of the Indian government's expenses, most notably through public sector projects like dams.

IV

Hopefully, this article has made a convincing argument that examining the performance of state-owned banks in a systemic crisis relative to private sector banks that have access to a weaker set of government guarantees is not a sound basis of assessing the overall attractiveness of state presence in the financial sector. While there is always a justification for greater presence of government institutions in the financial sector (or greater extent of government intervention in a crisis), this is likely associated with the misfortune of crowding out the private sector in the long run unless the government intervention has a graceful exit attached to its tail. Without such exit, government bailouts—and investor and depositor anticipation of such bailouts—for PSBs have deep consequences on competitive forces in the financial sector, potentially shaping their long-run form, and always stacking the odds against the flourishing of private banks. The growth of state-owned banks in the recent crisis thus raises several questions.

How will the private sector banks in India respond to the emerging strength of PSBs? Will they undertake greater risks and leverage in order to compensate for their likely underperformance when the next crisis hits? Can India afford such a race to the bottom when over 70 percent of the banking assets are funded in a manner that effectively represent off balance sheet liabilities of the government? And this too, in times when India's fiscal deficit as a percentage of its GDP is among the highest in the world? Similarly, will the private insurance sector be able to grow with reasonable premiums for a deprived population from a starting point where they must compete with LIC's effective monopoly and national champion status? As UTI's "unwinding" shows, things work out well in the private sector when government lets go off its muscular grip.

While the macro-prudential policies of the RBI have stood India in good stead against the real-estate boom and inflation risks to date, they are likely to come under increasing strain if the private financial sector sees demographic dividends disappearing beyond a few years and feels the heat to compete aggressively with the public players. Will RBI's asset-level leverage restrictions such as loan-to-value (LTV) ratios be meaningful if banks are lending at subpar rates and housing prices—denominators in the LTV ratios—are heavily inflated? Would nonbank financial firms develop a world of "shadow banking" for the private financial sector to park risks away from scrutiny of the RBI? Would it not be better instead to create a level playing field where all financial firms are private, but are charged upfront for government guarantees in the form of deposit insurance (up to reasonable ceilings), allowed to branch freely in rural areas as growth demands, and subject to a clear resolution authority, so that both financial firms and their borrowers undergo creative destruction and are, thus, incentivized to innovate and experiment? With better-run banks and a faster-growing economy, the economy could be weaned off priority-sector norms in due course down to a de minimis level.

The Indian state-owned banks are right now in the best of health they have been. Since many of them are already publicly listed, the way to their privatization and creation of a level playing field in the financial sector is relatively straightforward and needs to be undertaken without any further adieu.

Comments and Discussion

Kenneth Kletzer
University of California, Santa Cruz

This interesting paper considers an important question for financial sector stability in India: Does the dominance of state ownership in banking actually raise rather than reduce the risk of financial crises? The paper begins by considering a recent assertion in policy circles in India that the movement of deposits away from private banks to public sector banks in 2008 was a flight from riskier to safer banks. The empirical results show that the flow of deposits into the public sector banks is not explained by the financial positions of these banks. This suggests that the flight to safety is a flight to deposit insurance rather than to more sound banks. Viral proceeds to a more general discussion of the perils of government ownership of financial institutions with lending mandates. He argues that public ownership brings with it implicit, if not explicit, subsidies to insurance of bank liabilities that increase financial instability and pose a crisis risk for India.

The empirical analysis in the paper counters the argument that state ownership of banks is desirable because the public sector banks choose asset portfolios that are less exposed to severe financial risks than do private sector banks. The regressions show that deposit growth in 2008 is positively correlated with the MES for 2007 for private banks alone, but that deposit growth across public sector banks is negatively correlated with their MES. These estimates seem pretty convincing. Viral also shows that this measure of ex ante downside risk does not show any systematic meaningful difference between private and public banks on the eve of the crisis. We can see that something other than just exposure to severe adverse shocks explains the deposit shift. Another suggestive piece is his observation that share prices fell substantially more for private banks than for public sector banks with similar ex ante tail risk.

The natural explanation for the shift in deposits in the crisis is the one favored in the paper: savers perceive that they are more likely to recover their deposits from an insolvent public sector bank than from an insolvent private bank. Public sector banks are privileged in government-financed bailouts so that the interpretation that the deposit shift was caused by distortions in guarantees of the banking sector is sensible.

The raw data implies that relative balance sheet risk does not explain the shift. However, I have some concerns about the regressions. The result that deposit inflows are positively related to MES for public sector banks is perverse. It suggests that deposit growth may be determined by additional bank-specific characteristics or that the measure of risk is missing something about the relative security of deposits. A different measure of downside tail risk or the inclusion of more conditioning variables could change the coefficient sign and be more informative of how the risk of bank failure affects depositors' choices. If we accept that guarantees and prospective bailouts explain the flows, then we should observe either a negative or a null relationship between flows to public sector banks and the balance sheet risk of those banks. The significant positive correlation for the public sector banks does not negate this conclusion, but it does not inform it either. For example, bank size may matter to depositors and matter more for public sector banks because they are guaranteed. I think it is worth including additional regressors to raise our confidence in the regressions and in using the MES to measure the risk that matters to depositors.

It would be interesting to see how the relative growth in deposits is distributed between domestic and foreign depositors. Breaking down foreign resident deposits between nonresident Indian and other foreign sources might be interesting as well. The global flight to safety in the financial crisis is credited with the fall in US treasury yields from the middle of 2007 through 2008 and the coincident rise in yields for US corporate securities. India realized a reversal in capital flows between 2007 and 2008. A possible, and credible, explanation for the reduction in private bank deposits in 2008 is that foreign depositors favor private banks in India and reduced their holdings to purchase US treasuries or similar assets. A general global flight to safety led to a decline in foreign portfolio capital in India. This presumably was reflected in a reduction in bank deposits. Since domestic residents face capital controls for acquiring foreign securities and accounts, a reduction in deposits by foreign residents should not be matched by a proportionate reduction in domestic resident deposits (until the point that cash is safer than deposits). Under these conditions, a global flight to safety would lead to the observed contraction in deposit growth for private banks relative to public sector banks.

The observed increase in deposit growth for the public sector banks is small in comparison to the contraction in deposit growth for the private banks. This could be consistent with foreign and domestic depositors being distributed differently across private and public sector banks. Households and small enterprises probably favor public sector banks and among these, the ones with more branches. Without access to foreign asset markets, these deposits would continue to grow with domestic income. In the global flight to safety story I am telling, the shift away from private banks would be correlated with the riskiness of the private banks. Any exodus of foreign deposits from the public sector banks would also be correlated with the riskiness of deposits in those banks inclusive of the effects of government guarantees. If foreign residents hold only a small fraction of total deposits in the public sector banks, then the data may not reveal the dependence of foreign deposits on the insurance advantage of state ownership. The distribution of foreign and domestic depositors across private and public sector banks could be determined by government guarantees. Domestic depositors could seek those guarantees while foreign depositors seek higher marginal returns. But, the change in the growth rate of deposits in public sector banks would not show the effects of the difference in guarantees because of the impact of capital controls.

I think this paper makes a valuable contribution in refuting the argument that the shift in deposit growth away from private sector banks in 2008 evidenced less exposure of public sector banks to systemic risk than of private banks. My comments thus far are suggestions for looking more closely at the changes in bank deposit growth and discussing the effect of foreign capital flows on these changes in the crisis. Additional support for the hypothesis that guarantees of state banks drive the shift in deposit growth might be found by looking for differences in the responses to the crisis by depositors who are not protected to the same degree by government guarantees. All commercial banks, including branches of foreign banks are required to insure deposits up to one lakh rupees in size through the Deposit Insurance and Credit Guarantee Corporation. Some deposits are excluded from this regulation (e.g., deposits received outside India and interbank deposits). These could be more sensitive to the guarantees given to public sector banks than other deposits. The disaggregation might be useful for uncovering evidence that guarantees matter.

I now turn to the more expansive part of the paper which concerns the implications of state ownership of banks and of differential government guarantees for financial stability and the allocation of resources. My primary criticism of this very interesting discussion is that it is a big step from showing that the public sector banks are no less exposed to crisis risk than are private banks, and enjoy privileged government financial support to whether public sector banking is poor policy. The arguments in favor of privatization could benefit from more detail and nuance.

The most important topic, however, is that government guarantees of banks raise exposure to financial crises. We have seen the effects of explicit and implicit guarantees of banks following capital account liberalizations in emerging market economies repeatedly, and these are unpleasant. Such guarantees implicitly subsidize capital inflows and risk taking. In the absence of full capital account liberalization, the guarantees still distort the allocation of credit because bank owners and management do not internalize downside risks, but realize benefits from good outcomes. In the case of public ownership, we have examples of directed lending with public subsidization of lending and risk taking that result in boom–bust cycles.

Viral does a nice job explaining how guarantees with lending mandates can lead to overinvestment in socially less valuable capital and asset prices that reflect the private value of guarantees rather than the social value of investments. Since the resources of the government (or the will of its taxpayers) are ultimately bounded, the insurance indemnity cannot grow forever. Eventually, the market value of the claims exhausts investors' perceived value of the public sector guarantee, a crisis ensues and a bailout is demanded.

In this paper, Viral compares state ownership of banking in India to government-supported mortgage lending in the US. In other articles, he has elaborated how the mortgage GSEs in the US contributed to unsustainable investment in residential construction, loans to borrowers who could not service their mortgages and a housing price bubble. Fannie Mae and Freddie Mac are publicly held corporations that enjoy implicit government guarantees and are directed to lend in the interest of social goals. Without sufficient regulatory restraint, this combination encouraged risk taking by the GSEs and by private lenders who anticipated that the GSEs would intervene in the mortgage security market. This is an important story for understanding the financial crisis, although its centrality to the crisis is controversial. Viral's point in this paper is that Fannie Mae and Freddie Mac are examples of what directed lending with government backing can lead to.

The analogy to public ownership of banking in India is the combination of explicit government guarantees, directed lending, and public equity. The implication is that because the US mortgage GSEs were at the center of the housing boom and mortgage securities market that a similar institutional structure of public sector banks in India could also lead to financial crisis. I think that the problem of state ownership and directed lending for financial stability in India needs to be explained and articulated in terms of the characteristics of Indian banking and finance. The similarities in institutions are broad and the details of regulation and mandates matter. Whether a guarantee leads to excessive insurance liabilities depends on regulation of the activities of the insured. For example, the Housing Regulatory Reform Act of 2005 (US Senate 190) failed to pass. We cannot know if it could have averted the US crisis, but the arguments for it were on point. These concerned illusionary profits in mortgage lending and the stability of the mortgage GSEs. From this perspective, government ownership can inhibit reforms and prudent regulation. The point that privatization without subsidies and correctly priced deposit insurance is a better idea is well taken.

My last comment is that the paper casually states that public ownership of banks in India hampers productive lending to agriculture. There is a literature on how state and private banks lend in India and whether firms are constrained in access to loans. For example, Banerjee, Cole, and Duflo (2003) find a large positive difference between the marginal productivity of loans and bank lending rates. They also find that banks reissue loans at constant nominal values and show empirical evidence that firms are credit-constrained. Despite mandates, both private and public sector banks may not be allocating resources to productive rural investments in agriculture. In his paper, Viral mentions that private banks are inhibited from opening rural branches. The policy goes the other direction: a condition of opening new branches in already served (e.g., urban) markets is the addition of branches in markets without a bank. Viral's suggestion that rural areas and agriculture could benefit from private banking investment is supported by Burgess and Pande (2005). They study the effects of the branch expansion policies of the RBI that were in force between 1977 and 1990 and find that the expansion of banking into rural areas is associated with poverty reduction and diversification of rural investment.

Banks in India reveal a preference for investing in government debt and this tendency is more pronounced in public sector banks. The term "lazy banking" is commonly used at conferences about banking in India. This tendency probably makes the public sector banks less vulnerable to financial crises, but at the cost of lending usefully. I think the warning Viral is raising applies to the prospects for further financial liberalization quite well. In a more competitive environment, banks that enjoy free and expansive guarantees will accumulate risks. Moral hazard could be a big problem for lazy banks that start learning to lend. In this way, I think the analogy to the US mortgage GSEs and housing crisis is a good choice and an important point. For many years, everyone was complacent about these lenders and their mandates. In the 1990s, they were requested to enable more households to own homes as financial market deregulation changed the market environment.

The warning I hear in Viral's paper is to beware of government ownership that has looked innocuous in the past as financial reforms are made.

Urjit R. Patel

Brookings Institution

My comments as discussant are in six parts. After summarizing the main findings, I proffer alternative explanations for the author's main conclusions. Subsequently, I comment on and question the efficacy of the central applied "tool" of the paper, namely, MES. Then I raise aspects of Indian banking that should have been examined and critically explored to impart an institutionally richer texture to the paper; implication of this for national debt sustainability calculus is highlighted. Later I comment on the analogy to the US GSEs that the author makes. Finally, I conclude.

A. The proximate findings of the paper are:

    1. Indian PSBs came out ahead compared to the private banks during the global crisis in 2008. There was, in some sense, a domestic flight to safety at least in terms of garnering deposits in favor of the PSBs, and this translated into: (*a*) relatively better equity valuation for PSBs vis-à-vis private banks (and the "relative" is important because equity valuation of *all* banks in the sample declined); and (*b*) lower CDS spread for SBI compared to ICICI Bank, although CDS spiked up (and stayed high) for both.
    2. The share of private banks in the sector declined for possibly the first time in two decades after gradually rising up to over 70 percent.

B. The paper asserts that the above outcomes are undesirable and are wholly attributable to government guarantees to PSBs (i.e., driven by moral hazard). The paper also finds that among the private banks, those with higher risk had lower deposit growth and I think this is consistent with the model that the author has in mind (although it is not formally sketched out).

   But since government guarantees are binary in nature, why should there be dispersion around a positive trend between risk and deposit growth among the PSBs? This latter finding is not explained in the paper. Since the model suggests that government guarantees drive higher deposit growth, individual characteristics of banks are largely irrelevant in the model as is presented in the paper.

There is another explanation for the main findings mentioned earlier. The Indian depositor essentially behaved like an equity investor. They reallocated their deposit portfolios away from private banks to PSBs. Why? A plausible answer is that Indian private banks may have suffered on account of reflected infamy. Around the crisis period, they were viewed as too avaricious for Indian depositors' taste. It was thought that much like their counterparts elsewhere—New York, London, and Reykjavik—who were role models for Indian private banks, they too may have been prone to misbehavior. It was probably also perceived that private banks were more exposed to the external environment than they were admitting to, and certainly the foreign banks based in India had imbibed bad habits like underreporting unhedged risks on their balance sheets. This driver of flight to safety is distinct from government guarantees to PSBs.

Is the decline in the market share of private banks permanent? The deposit reallocation in 2008 and 2009 was on account of the global crisis, but this has now dissipated. More recent data needs to be examined to assess whether a relative decline in the share of private banks compared to PSBs is a persistent and enduring feature, or, whether it was a temporary phenomenon.

Were the PSBs able to attract more deposits because they offered high deposit rates in 2008–09? It could be that they were more desperate than the private banks during the crisis despite ample liquidity support; bank-specific interest rate data on fixed deposits may shed some light in this context.

C. Turning to the MES, the outcome of the flight to safety to PSBs is observed in spite of PSBs, on average, being ex ante more vulnerable compared to private counterparts as measured by the MES.

The specific case study of SBI and ICICI banks is a good one because these are the largest banks in India—the former is government owned and the latter is privately owned. SBI and ICICI banks had the same MES prior to the crisis in January 2008, but divergent performance thereafter with the former gaining post crisis.

Two dimensions need to be kept in mind for the MES exercise:

- Behavior of the measure beyond the paper's sample period.
- Diagnostics to determine statistical confidence in the MES measure.

It may be informative to extend the analysis to at least one full fiscal year beyond the crisis to determine persistence. It could be that private banks have started to regain market share and we need to be aware that the values of average MES have changed markedly over this period. Using Bloomberg data I calculated the MES measure for 42 months starting January 2008 to June 2011 (see Figure 8). It is obvious how much the MES has varied and, in fact, by June 2011, ICICI Bank had lower risk compared to SBI. I used these two banks only as examples, and it could well be that we would find something different among other (bilateral) comparisons from the sample of banks used in the paper.

The MES measure may also be very sensitive. The very short trading horizons at which equity markets operate, as also the possibility of return to normality post crisis, suggest that the MES has to be subjected to some diagnostics. At first pass, standard errors would be standard (and useful) in this regard. It could be that the MSEs of individual private and public banks may fall in the standard error band and, therefore, point comparisons at specific dates may be misleading for drawing conclusions. I plotted the standard errors of the MES for the same 42 months and it is clear that there is a large degree of oscillation (Figure 9).

FIGURE 8. Monthly MES, January 2008–June 2011

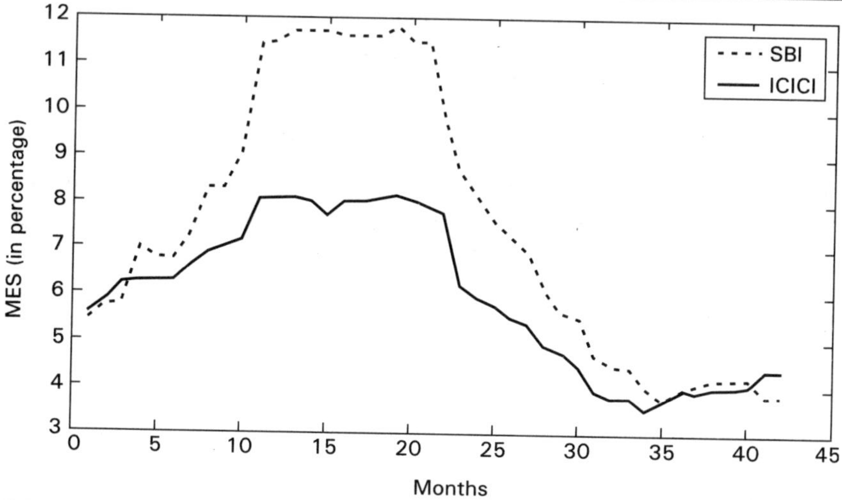

Source: Bloomberg data.

FIGURE 9. Standard Errors of MES

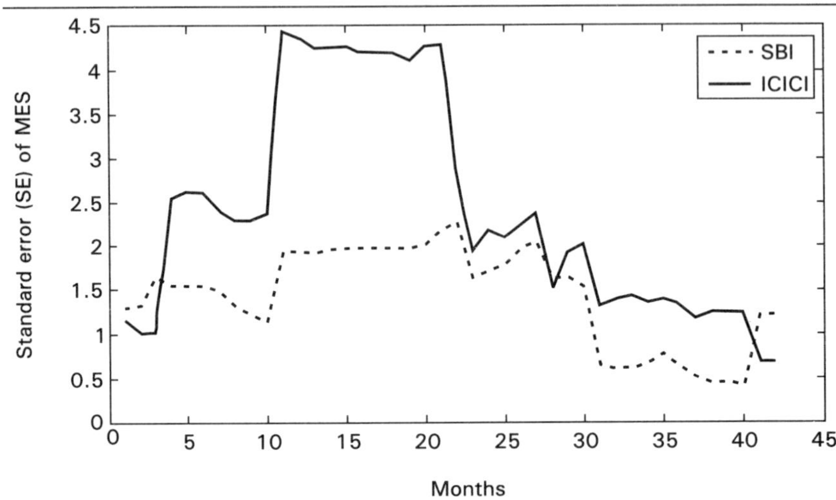

Source: Bloomberg data.

D. In terms of details regarding the banking system, the paper could benefit by looking at another set of variables, for which data is available, to gauge and examine the underlying forward-looking risk profiles of banks. For example, implications emanating from government ownership related to commercial and operational functions of the banks: these are related to outcomes for nonperforming assets, the asymmetry in provisioning policy that we see between public and private sector banks, differential net interest margins that banks earn, etc. This kind of forensic data would have given the paper's theme not only a highly relevant broader perspective, but would also have enhanced the case that the author sought to make.

How credible are (blanket) guarantees (i.e., when they are given to one and all)? In India, even private sector banks when they have got into problems have brought about early and decisive intervention by the authorities, especially the RBI; assurances have been made that depositors will be protected. Take the case of Global Trust Bank, one of the star private sector banks, which was established, operated/expanded and wound up—all within 10 years! The bank failed and was taken over by a PSB, the Oriental Bank of Commerce, in a deal shepherded by the RBI. ICICI Bank, at the receiving end of, I think unfounded rumors in 2003 and 2008, warranted strong, credible, and fairly prompt guarantees by the RBI that all depositors will be protected. I think when we look at the credibility of the guarantee to depositors, the ownership per se is not important. Recent crisis in the US, the UK, and the Eurozone has amply demonstrated that entities that are too big, too complex, too interconnected, or too politically connected to fail are always bailed out regardless of ownership, and private debts then become public debts. To then assess how credible this guarantee is (i.e., whether the state actually has the requisite deep pockets), the government's debt sustainability calculus has to be rescrutinized by encompassing the government's assumption of contingent liability for the sector. Ireland is a recent basket case example where the guarantee to the private banks has overwhelmed the country's GDP, let alone the sector.

E. The analogy with US GSEs needs to be nuanced and a fair bit of caution is warranted in this context. The risk exposure of Fannie Mae and Freddie Mac is not only to one sector; it is, in fact to one subsector within a single sector—housing within real estate—and none of the Indian PSBs are anywhere near those kinds of sector risk concentrations. If anything, Indian banks are fairly diversified on the asset side.

   There is another question in the US context, which requires consideration, but which is wholly ignored in the paper. Was it solely the credit enhancement role facilitated implicitly by the government that led to the ultimately unsustainable boom in the US housing market? To be honest, there is a lot of blame to go around, namely, long-standing fiscal incentives for home buyers; the nonrecourse nature of home loans in the US; the overhang of loose monetary policy by Greenspan and his able deputy Bernanke with substantial deviation from Taylor's rule for long periods of time (see, e.g., Taylor, 2009); and the role of (perverse) incentives within Wall Street. In other words, there are broader drivers, deeper explanations, and qualitatively important reasons, rather than just the GSEs guaranteed by Uncle Sam for the housing finance-inspired crisis.

F. On the policy side, the paper argues for privatization of Indian PSBs and other government-owned intermediaries (LIC is mentioned in the paper). It is apposite to ask how closing the safe option for Indian savers be welfare enhancing. Can it impact household financial savings and India's overall savings ratio? It is clearly an issue that needs to be looked at more closely than is the case in the paper, especially if there is a flight to safety cause rather than government guarantees. My conclusion to the paper is that many of us would agree that financial intermediation in India will benefit from privatization and reading this paper does not dilute this assessment, but unfortunately this paper does not help to reinforce it either.

General Discussion

Renu Kohli questioned the interpretation that the private–public distinction for deposit growth was the result of a government guarantee since past policies seem to be to never allow any bank, public or private, to fail. She thought other factors could account for the divergence of deposit growth. For example, private banks may have had a greater reliance on short-term deposits which could be expected to be more volatile. She thought it would be useful to look at deposit rates in public and private banks to determine if they responded differently.

Dilip Mookherjee said that in terms of the main issue of whether banks ought to be privatized, the evidence in the paper is inconclusive because that issue would involve a much broader set of considerations than are discussed in the paper. The results of the empirical analysis should be expected: when you have a system where one sector has more of a government guarantee, those that have the guarantees take more risks, and when there is a crisis, there is a flight to safety and people move into that sector because of the guarantees, even though they took more risk. That does not necessarily speak to the broader question of whether banks ought to be privatized.

Devesh Kapoor supported a point made in Urjit Patels' comments, which was that the flight to safety to public banks might not have been just because of the guarantees that those banks hold, but also because of how the depositors perceive private banks versus public sector banks. He mentioned a survey done by the Raghurajan Committee which asked about public trust in different types of banks, and it found that while only 5–10 percent of the public trusted private banks, 80 percent trusted PSBs.

Abhijit Banerjee pointed out that it is hard to think that the private sector is very wise in its risk-taking decisions, and conversely, although government banks might be taking on a little more risk because of the guarantees, that might be a good thing because otherwise the people who run them have very little incentive to take on much risk. He also mentioned that there are really two different sources of risk, one being the portfolio, which during normal times might be less for private firms who exert more effort to select the better borrowers. The other is the risk that kicks in during a crisis, which is the risk of the bank going under. Therefore, the change in risk profile is an important and interesting feature to track.

References

Acharya, Viral, Lasse H. Pedersen, Thomas Philippon, and Matthew Richardson. 2010. "Measuring Systemic Risk," Working Paper, New York University Stern School of Business.

Acharya, Viral and Nirupama Kulkarni. 2012. "State Ownership and Systemic Risk: Evidence from the Indian Financial Sector during 2007–09," Working Paper, New York University Stern School of Business.

Acharya, Viral, Stijn van Nieuwerburgh, Matthew Richardson, and Lawrence White. 2011. *Guaranteed to Fail: Fannie, Freddie and the Debacle of Mortgage Finance.* NOIDA, UP: HarperCollins India.

Banerjee, Abhijit V., Shawn Cole, and Esther Duflo. 2004. "Bank Financing in India," in W. Tseng and D. Cowen (eds), *India's and China's Recent Experience with Reform and Growth*, pp. 138–57. Hampshire, UK: IMF and Palgrave-Macmillan.

Berle, Adolf and Gardiner Means. 1932. *The Modern Corporation and Private Property*. New York: Harcourt, Brace & World, Inc.

Burgess, Robin and Rohini Pande. 2005. "Do Rural Banks Matter? Evidence from the Indian Social Banking Experiment," *American Economic Review*, 95 (3; June): 780–95 [American Economic Association].

Chandrashekhar, C.P. 2008. "The Importance of Public Banking," *Frontline*, 25 (25; December).

Cole, Shawn. 2009. "Fixing Market Failures or Fixing Elections? Agricultural Credit in India," *American Economic Journal* (*Applied Economics*), 1 (1): 219–50.

Economic Times. 2009. "Deposits with SBI Zoom Past Lehman Collapse," April 7, Mumbai, India.

Gokarn, Subir. 2011. "Inflation Is Still a Real Threat to Global Growth," *Financial Times*, March 29.

Government of India. 2010. "Apex Level Financial Stability and Development Council to Set Up for Maintaining Financial Stability," Press release ID: 58428, Press Information Bureau, Government of India. Available at http://pib.nic.in/newsite/AdvSearch.aspx

RBI. 2009–10. "Trend and Progress of Banking in India 2009–2010," Annual Publication, Reserve Bank of India. Available at RBI official website.

Subbarao, Duvvuri. 2009. Speech delivered at the symposium on "The Global Economic Crisis and Challenges for the Asian Economy in a Changing World," Institute for International Monetary Affairs, Tokyo.

Wharton. 2009. "As Foreign Banks Detour, Public Banks Forge Ahead," India Knowledge@Wharton.

World Bank. 2009. "International Bank for Reconstruction and Development, Program document for a proposed loan in the amount of US$2.0 billion to the Republic of India for a Banking Sector Support Loan."

AJAY MAHAL
Monash University

VICTORIA FAN
Center for Global Development

Expanding Health Coverage for Indians: An Assessment of the Policy Challenge*

ABSTRACT As India's policymakers consider how best to reform its health system, the goal of "universal health coverage" (UHC) and the steps required to achieve it have garnered considerable attention. In this paper we inquire into some of the conceptual and implementation issues relevant for the expansion of UHC. We revisit the case for government intervention in the health sector and inquire whether the case for UHC can be justified on grounds other than as "human rights." We assess implications of resource constraints and examine specific questions of implementation pertinent for expanding health coverage. Finally, we assess the resource implications of a UHC policy, and an alternative that focuses on just the poor.

Keywords: *Universal Coverage, Insurance, Right to Health, Payment, Public Services, Financing, Risk Protection*

JEL Classification: *H4, I1, I13, I18*

Introduction

As India's policymakers consider how best to reform its health system, the goal of UHC and the steps required to achieve it have garnered considerable attention. The recent report of the High Level Expert Group (HLEG) set up by the Planning Commission, henceforth referred to as the HLEG report, supports a strategy of achieving universal coverage (HLEG, 2011). Achieving universal coverage in India was also the subject

* We are grateful to participants at the India Policy Forum workshop held in New Delhi on July 2011 for their comments and to Arvind Panagariya for critical commentary on the previous draft that helped greatly improve this paper. We thank Rachel Silverman for helpful research assistance. We alone are responsible for any errors that remain.

of an advocacy piece published earlier in the *Lancet* by some of the authors of the HLEG report (Reddy et al., 2011)which proposed that *all* Indians ought to be able to access treatment to a "full range" of health conditions. These developments are seen as an reaffirmation of the vision outlined in the Bhore Committee Report of 1946, the Directive Principles of India's Constitution, and various international declarations and covenants to which India is a signatory, including the "Universal Declaration of Human Rights" (HLEG, 2011; Rao, 2012). And they are underlined by recent legislative initiatives, of which one striking illustration is the National Health Bill under consideration in Parliament. The Bill aims to "... provide for protection and fulfillment of rights in relation to health and well-being, health equity and justice" (Ministry of Health and Family Welfare, 2009: 6).

The advocacy of universal coverage accompanies multiple major recent efforts intended to improve service quality and affordability of health services in India. These include the central government-funded National Rural Health Mission (NRHM), initiated in 2005 and aimed at improving service quality of primary and secondary care, and the Rashtriya Swasthya Bima Yojana (RSBY), a health insurance scheme for the poor launched in 2007 that provides (secondary) hospitalization coverage and is funded in a 3:1 ratio by the central and state governments (Palacios, 2010; Swarup, 2011). Health insurance schemes have also been launched by various state governments, such as *Aarogyasri* in Andhra Pradesh, the *Vajpayee* Scheme in Karnataka, and the Chief Minister's Elaborate Medical Insurance Scheme (previously known as *Kalaignar*) in Tamil Nadu, although these are aimed at tertiary care and provide more generous coverage than RSBY.

International Trends

The developments in India parallel increased recent prominence of the topic of UHC in low- and middle-income countries as a matter of policy and academic interest. Universal coverage was the main theme of a recent World Health Report (WHO, 2010) and of two major recent conferences, the Global Symposium on Health Systems Research held in Montreux, Switzerland in November 2010, and Prince Mahidol Awards Conference, held in Bangkok in January 2012, both including donor organizations, leading health researchers, and policymakers.

Several countries in Asia, Latin America, and Africa have drawn international attention for their health sector reforms aimed at universal coverage. Among Asian countries, Thailand, China, and Indonesia stand put as prominent recent examples in this direction. Thailand introduced a tax-financed insurance scheme in 2001 that extended financial risk protection to an additional 18.5 million population, thus bringing almost its entire population under insurance cover (Hughes and Leethongdee, 2007). Indonesia launched a health insurance scheme for the poor in 2004, currently covering more than70 million Indonesians, with the ultimate objective of bringing all Indonesians (including those who are currently enrolled in social insurance schemes for formal sector workers) under one cover (Rokx et al., 2009). The Chinese government embarked upon an ambitious program to extend health coverage to its entire rural population under the New Cooperative Medical Scheme. Currently this coverage extends to more than 90 percent of China's rural population (Yip and Hsiao, 2009; Tang, 2011). A separate urban insurance scheme covers individuals in China's urban areas (Liu, 2002).

In Latin America, the Mexican government expanded coverage to nearly 30 million previously uninsured individuals under the Seguro Popular health insurance program with the aim of reaching the (additional) remaining 18 million uninsured by the end of 2011 (Frenk et al., 2006). In Africa, Ghana introduced National Health Insurance Scheme in 2003, which has expanded in six years to cover nearly 55 percent of its total population (World Bank, 2011), and South Africa shortly plans to launch a National Health Insurance scheme of its own.

Key Issues for India

These country examples and the developments in India reflect two central observations about international trends in health spending. First, total health spending per capita is continuously rising. Second, pre-paid health spending (through government health subsidies or other insuring mechanisms) increases as a share of total health expenditure over time, while the share of out-of-pocket spending decreases (Fan and Savedoff, 2012). Together these two features are explained by a number of factors—rising incomes, a burden of disease that is shifting to chronic conditions that are expensive to treat, and higher demand for financial risk protection, thereby pushing governments to promote expanded levels of coverage. There is also some evidence that countries with multiple risk pools and mixed systems have higher costs because of weak cost control, when compared to countries with a single-payer system, either through a national health service (NHS), UK, or national health insurance model (Xu et al., 2012).

The push toward expanded coverage in India (and elsewhere) is not without concerns. On his popular World Bank blog, Adam Wagstaff (2010) notes that the usage of the term "universal health coverage" (UHC) may itself be confusing, given that it is often taken as "either-or" state (i.e., a country either has universal coverage, or it does not), instead of visualizing countries as lying on a coverage continuum—something that other authors have also highlighted (Wagstaff, 2010; Scheil-Adlung and Bonnet, 2011; WHO, 2010). Wagstaff suggests that policy steps intended toward achieving universal coverage have a mixed record in achieving desired health outcomes in both developed and developing countries, at least in terms of improving equity in health care use, although its impacts on improving financial risk protection are more substantiated. That leaves open the question of whether other approaches are likely to be at least as effective in influencing outcomes of policy interest.

In the Indian context and despite its recent advocacy, the justification for a strategy aimed at UHC and its implementation are inadequately addressed in policy analyses. The HLEG report envisages a National Health Package accessible to all for free and financed primarily by public spending, expected to increase from about 1.2 percent of GDP at present to 3 percent of GDP by 2022. However, the report does not provide much information on the benefits that would be provided under the proposed UHC or how they would change over time; or crucially, provide a justification of the desirability of a package *accessible equally to all* right at the beginning of the implementation process, other than as a basic right. Nor is the basis for its cost estimates for UHC sufficiently well explained. A recent article argues that the HLEG report falls short on many specifics, including adequately defining the health services to be covered in light of resource constraints, addressing issues of incentives relevant to both public and private providers, and tackling the broader governance and capacity issues that underpin a large-scale expansion of health services. It concludes that the report "... severely falls short in providing a blueprint on what, how, and at what cost, and with what trade-offs, making it difficult to translate the ideas into operational strategies ..." (Rao, 2012, p. 15).

Goal of This Paper

The purpose of this paper is to inquire into some of the conceptual and implementation issues relevant to the expansion of health coverage for Indians, including the proposals of the HLEG report related to universal coverage. In particular, we revisit the case for government intervention in the health sector and inquire whether the case for UHC can be justified on grounds other than as a "human right." We assess the implications of resource constraints for health sector policies aimed at expanding coverage in the Indian context. We also examine specific questions of implementation pertinent for expanding coverage: What are the appropriate principles in designing a benefits package? How should health care be paid for? How should prevention and health promotion activities be provided and financed? How can the supply of good-quality health services be ensured? In doing so, the paper will draw upon recent experiences from within Asia and elsewhere. Finally, we assess the resource implications of a policy intended to achieve universal coverage, and an alternative that focuses on just the poor.

Is Universal Coverage Appropriate for India?

We first formulate a conceptual framework to examine the case for government intervention in the health sector. We then draw implications for (*a*) the health services to be covered and (*b*) specific population subgroups of interest, such as the poor (and near-poor), females, and children.

When Should Governments Intervene in the Health Sector and How Should Public Funds Be Used?

A simple framework due originally to Ehrlich and Becker (1972; also Baeza and Packard, 2006) is useful in thinking about what to cover and whom to cover from public funds (for the moment we will abstract from access and quality issues).

Consider an individual who faces a risk of becoming ill, and there is a chance of a bad outcome consisting of both health and financial elements depending on the care that is sought and/or available. When faced with financial (or health) risks on account of illness, risk-averse individuals/households can respond in at least four different ways. First, if available, they can voluntarily purchase insurance policies in the market enabling them to obtain needed health services. Alternatively, they can self-insure themselves up to a point, by either increasing savings or relying on community mechanisms that involve households supporting/lending to each other during times of need. If the likely magnitude of losses is high and formal insurance is expensive, households may be able to reduce (or postpone) the risk of loss through preventive action. Finally, there is the possibility that they cope with the loss itself, which could involve foregoing care, or paying for treatment. Foregoing care could be either intentional or unintentional; unintentionally foregoing care may arise from refusal to treat by providers against individuals who appear to be unable to pay for care.

The above framework can be extended to children living in households with some modifications, given their lack of decision-making authority. First, adults may not be interested in the well-being of children that they are a guardian of, or they may be interested in a subset of their offspring, such as male children. Thus, if a sick child is left untreated, there may be no treatment costs for adults, but there may still be longer-term adverse implications for children in terms of cognitive skills, schooling attainment, and productivity when they do join the labor market. These effects may be especially significant for female children in some societies. Even if adults in the household are concerned about the well-being of their children, they may lack the vision to address these risks adequately. In these circumstances, self-insurance, community support systems, or voluntary purchase of insurance are not particularly relevant as mechanisms to protect children against the financial risks of ill health.

When losses from the risks in question are small and infrequent, simple coping behavior should suffice, and in any case, formal insurance coverage in such cases is likely to involve relatively high administrative costs. However, if adverse events are low cost but frequent, prevention activities that lower the probability of adverse events, possibly in combination with self-insurance and coping, may be the optimal strategy (for an exception, see later). In both of these circumstances, it is unlikely that the benefits of formal insurance mechanisms would be large enough to make it worthwhile for an individual to obtain insurance.

With respect to losses of significance, insurance is the preferred option. For rare events involving potentially large losses, individuals and small communities are unlikely to be able to self-insure, or cope adequately. Large losses could also occur with high frequency (e.g., chronic conditions). The response in terms of insurance coverage should be for the market to create larger risk pools for purposes of insurance, and cover populations across multiple time periods (for chronic conditions), or across geographic regions. However, it is well known that health insurance markets are inefficient in providing the needed coverage and poor risks may be left underinsured (Newhouse, 1996). Because "large" is relative, this category should be broadened to include frequent small losses for individuals who are living close to impoverishment levels, and for whom every small "shock" is potentially catastrophic. One example is the evidence from developing countries on high out-of-pocket expenditures on medicines by households (Garg and Karan, 2009).

Preventive actions are crucial in almost all segments of the cases considered: either lowering the likelihood of a specific condition, or delaying the losses associated with it. Many preventive activities with substantial externalities are also undersupplied if left solely to households and private suppliers. These include not only vaccination and actions toward preventing communicable disease, but also actions like providing information on healthy eating habits, exercise, and hand washing. Preventive interventions are of particular importance to the long-term health and well-being of children and the incentives for household action here is even less for reasons mentioned earlier. Individuals also happen to be poor judges of their own future health, and they may be uninformed about their own benefits from prevention activity, such as from not smoking, or regular medical check-ups, and they may not be adequately forward-thinking, putting off exercise or adopting healthy eating habits. Prevention actions can also involve patients seeking care early for an ostensibly minor illness, thereby preventing potentially serious conditions later. In this category we could also include victims of accidents (whether rich or poor) who are unable to make a decision on their own behalf.

Implications for Design of Coverage

The preceding discussion points to the following efficiency argument for public intervention aimed at raising social welfare. Illness-related risks justify supporting coverage for health conditions that impose large financial risks on households, such as low frequency and high cost events and high frequency and high cost (usually chronic) conditions, such as cancer and diabetes. The latter, as mentioned earlier, could be covered via insurance policies that span multiple time periods. The well-known problems of voluntary insurance markets imply that any such insurance would require government regulatory intervention at the least, including some degree of compulsion for the consumer to participate in such insurance. In addition, governments should intervene to promote preventive actions for communicable and non-communicable conditions for at least two other reasons—disease externalities as well as behavioral concerns relating to individual lack of self-control and efficacy to prevent future health risks (e.g., relating to hyperbolic discounting). Our discussion also supports government intervening to help children access needed health services and preventive services, irrespective of economic status.

A second set of arguments for public intervention stem from an equity justification.These include concerns about gender equity, which can justify public intervention for maternal and child health services. In addition, low incomes impose additional constraints on what households can do, in terms of being able to purchase insurance (if available), or being able to cope with frequent albeit small expenditures. This means that government funding may be required to support insurance coverage for the poor, as well as for prioritizing some categories of prevention and health promotion programs for them.

There are essentially two different ways to achieve these goals. The *first* relies on separate insurance pools for a benefits package on curative care, supplemented by publicly funded health services for children and preventive and health promotion services accessible for all. Both Thailand and Mexico are examples of such a model, and health systems of other countries in the region, such as Indonesia, have these features. In Mexico, there is compulsory contributory coverage for formal sector workers and civil servants in separate pools. For the rest, primarily informal sector workers and others, there is the pool known as *Seguro Popular*, which has means tested and highly subsidized premiums, but with voluntary participation. In addition, there is one pool for infrastructure and human resource development, and another pool for financing prevention and promotion. In Thailand, similarly, there are separate pools for civil servants, social security workers with compulsory participation, and zero premiums for the rest in a separate pool. As in Mexico, there are additional pools for infrastructure (financed out of general revenues) and prevention/promotion.

In Mexico, public subsidies are transferred to *Seguro Popular* through three mechanisms. First, general revenues are used to support some of the subsidies, from both the central and state governments. Second, fund allocations to Mexican provinces which have their separate pools, are inversely related to their level of development; and finally, the (small) portion of the premium contributed by the enrollees into *Seguro Popular* is proxy means tested. In Thailand, 70 percent of the population (all informal sector workers and others not covered by social insurance) is covered under the universal coverage scheme—mostly rural and agrarian—does not pay any premium for the scheme which is financed from general revenues, a form of equity pooling. The key idea is that members of the formal sector and civil servant pools pay higher contributions (and fewer subsidies), albeit they may receive better benefits coverage.

There is, of course, another possibility that in consistent with having separate pools for the rich and poor. Under this option, the government could choose to limit its subsidies for a fund that provides coverage for specific population subgroups (such as the poor) and leaves the rest of the population to whatever they can obtain in the private sector. This model is likely to result in considerable welfare losses in the presence of insurance market failures and unexploited economies of scale that arise when individuals rely on out-of-pocket payments without a central purchaser to act on their behalf for health services. This seems unattractive as a policy option even if government subsidies are intended solely for the less well-off.

The *second approach* is to rely on common benefits in a publicly funded system, usually with little or no charges to users of health services. One version of the model relies mainly on public provision of health services, as in Sri Lanka and Malaysia (as also India to an extent). A second version involves combining all contributions, including subsidies, into one single pool (or separate pools for curative care and prevention) with options to purchase services from public and private providers, as suggested by Reddy et al. (2011) and also the HLEG report. The latter version of a publicly funded system recognizes both public and private providers, and under the right conditions, competition between private and public health sectors could be used to help to address long-standing problems of poor quality in the public sector.

Because health care services, including preventive care and health promotion activities, are financed out of general revenues to which the relatively wealthier populations contribute more under the second approach; it is sometimes argued that richer populations are paying more than their costs to the system, i.e., cross-subsidizing the less well-off participants of the system. However, this is not a certainty as we argue below, and additional revenues may need to be raised. Australia is an example of a model that imposes additional (earmarked) taxes on the richer individuals who choose to rely on its publicly funded services. Specifically, individuals who choose not to buy private hospital cover to obtain services in the private sector (separate from the government health-care services that are available to everyone for free) have to pay an extra tax of 1 percent of their gross income on provided their income exceeds a specific threshold.

The above models are also consistent with the use of regulation to influence some types of preventive actions, such as the banning of smoking in public places, or rules relating to immunization status for admissions to schools and child-care facilities that are often seen in developed countries. However, developing countries are usually not in a position to implement regulations and India has a poor track record in this area. Strict imposition of regulations could also have unintended adverse consequences for other desirable goals, such as school attendance among children.

Universal Coverage and Resource Constraints

As the preceding discussion makes clear, both financing models described earlier—the single pool and multiple pool options—are capable of delivering risk protection for insurable conditions, as well as preventive care and health promotion. If we define "universal coverage" as a multidimensional concept involving access to a minimum desirable set of services (quality a given), a threshold for ease of physical access (distance, timing, appropriateness), and low financial burden (risk and impoverishing effects) as in Figure 1 (see also Scheil-Adlung and Bonnet, 2011; WHO, 2010, for similar definitions), then it is plausible that an outcome that meets UHC thresholds could be achieved by either of the two financing models. Universal coverage as defined here is perfectly consistent with reasonable objectives about efficiency and the well-being of the poor if resources are no object.

In the presence of resource constraints, as in India's situation, UHC is essentially an aspiration and both models are alternative financing approaches that can be used to get part of the way there. Given that significant differences in financial and locational access exist in India across socioeconomic status, the goal of attaining universal coverage at some point in the future almost certainly requires lowering inequalities in access and enabling the poor to achieve some minimum degree of access, outcomes which would

FIGURE 1. Understanding (Universal) Coverage

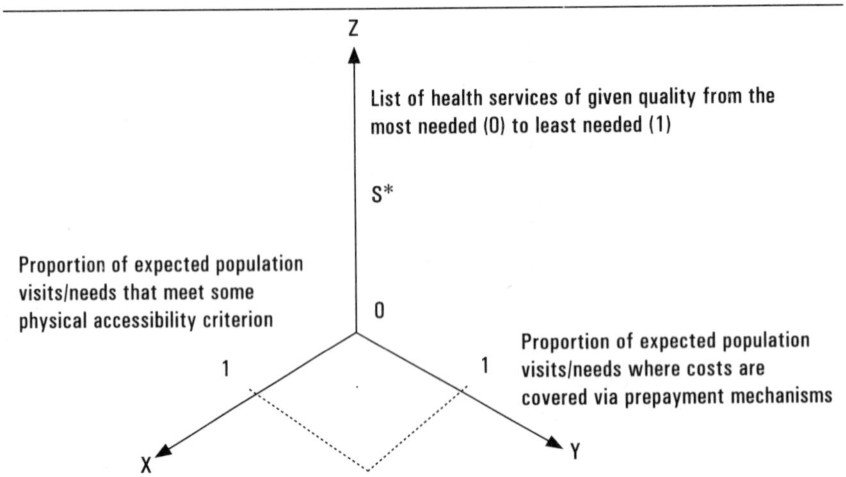

be desirable under reasonable normative principles. For instance, consider an inequality-averse or, in the extreme, a Rawlsian-type social welfare function. If health status, or protection against financial risk, is considered a major determinant of individual well-being, promoting social welfare will support the principle of enhancing the health (and health services) and financial risk protection of the less well-off relative to their better-off counterparts. Versions of the two pooling frameworks described earlier that require only the poor being subsidized are prima facie consistent with social goals, whether couched as a pathway to UHC or justified on the basis of some other welfare criteria.

The current drive to UHC is not purely aspiration though. Specifically, there is underlying legal enforceability, directed toward the government, of guaranteed access to some minimum set of services for *all* (World Health Assembly, 2005). This reflects the influence of the human rights field that emphasizes the "... *right* of everyone to the enjoyment of the highest *attainable* standard of physical and mental health ..." (International Covenant for Economic Social and Cultural Rights, Article 12.1). While there is considerable weight given to the idea that the realization of the "right" is conditional on available resources, there is no doubt that legal enforceability is critical (Rao, 2012; Scheil-Adlung and Bonnet, 2011; Wagstaff, 2010). In the HLEG report itself, there is the statement: "The foundation for UHC is a universal entitlement to comprehensive health security and an all-encompassing obligation on the part of the state to provide adequate food and nutrition, appropriate medical care ..., health related information and other contributors to good health" (HLEG, 2011, p. 3).

If legal enforceability of patient rights, e.g., enables the poor to better access a defined set of health services, then we see no conflict with the preceding discussion. The problem arises when legal rights to subsidized health services are extended to all, as envisaged in the HLEG report. While such a guarantee may be appropriate at a stage where a government is wealthy enough to support a wide range of services for large numbers of people across the socioeconomic divide, it is unhelpful when there are large existing inequities in population access to services in a setting with resource constraints. A good example of this situation is the public sector health services in India which, in principle, are accessible to all for free, but provide far greater benefits to better-off sections of society in a resource-constrained environment (Mahal, 2000). Unequal access was also observed in the Aarogyasri scheme for hospital insurance in Andhra Pradesh, with Scheduled Caste and Scheduled Tribe populations not benefiting from programs as much as other groups (Fan et al., 2011). It is fortunate that access to a minimum set of public sector-funded services is not yet defined as a legally enforceable "right" in India, otherwise poorer populations in India would be further disadvantaged by their relative lack of access to legal mechanisms compared to the nonpoor.

Guaranteeing everyone the same degree of access to (subsidized) services when resources are very limited usually will not serve the goal of enhanced equity, or help the less well-off achieve some minimum coverage threshold. With limited extra funds, governments will have to decide whether they wish to provide existing levels of subsidies to a larger beneficiary group (breadth of coverage), or a larger subsidy to a subgroup of needy beneficiaries (depth of coverage), or something in between. Nor can it be claimed that the rich pay disproportionately more by way of taxes into general revenues used to finance public sector health services (implying that access is *not equal* even if everyone has the same benefits). The rich tend to use more services (they live longer and are better able to access health care than the poor) so the actual degree of cross-subsidization across economic groups may end up being quite limited. That is, the focus of efforts to improve access should either be exclusively the poor, financed by general revenues; or if the rich also have guaranteed access to the same set of health-care services as the poor, they should pay substantially more, say via additional earmarked taxes, or user fees to allow for adequate levels of cross-subsidy. Separate pools for the rich and the poor, with the latter funded from general revenues as in Thailand and Mexico, and the former being based on employee/employer contributions may be another vehicle to do this (Suchonwanich, 2010). Once this is done, legal enforceability may even be desirable from the standpoint of program implementation.

Implementing Options for Coverage: Issues

Once the subgroups have been decided and the resource envelope broadly determined, multiple issues related to implementation of program coverage must be addressed. There is firstly a need for clarity on the services to be provided and the associated financing mechanisms (country examples discussed in the previous section shed light on some of these issues). Second, if the poor are designated as the beneficiaries of public subsidies, adequate mechanisms will be needed to identify this group. A third issue has to do with how health services are paid for. Finally, supply side issues have to be considered, so that the necessary services are available.

Which Services Should Be Provided?

Given resource constraints, and a social objective that favors reducing inequality, it makes little sense to subsidize better-off populations, other than for the purpose of influencing behaviors with significant spillovers, or when individuals are not in a position to judge their own future health implications of their current actions, such as preventive health check-ups. For the poor, insurance contributions will also need to be subsidized, whereas insurance for better-off groups may need regulatory support.

Should the coverage for financial protection include *all* conditions? Even apart from the issue of constrained resources there are good grounds for arguing against 100 percent coverage due to efficiency considerations, given that administrative costs (of processing multiple small claims, for instance) can be high. Concerns about moral hazard, both on the demand and the supply side of the health-care market, are also legitimate in this context. Disincentives for certain kinds of preventive and coping behavior may result, such as households not seeking their own cough and cold remedies and not pursuing preventive behavior, instead favoring a visit to a health-care provider. Individuals are usually not in a position to effectively judge the quality of care they receive and may err on the side of excess and ineffective care, a tendency facilitated by high levels of coverage and providers for whom financial returns are associated with the services they provide. When there are multiple pools for insurance, it is entirely plausible that better-off people paying for their own insurance coverage enjoy more generous benefits than beneficiaries of a pool subsidized from public funds. Indeed, in an unregulated competitive market with voluntary purchase of insurance, the benefits cover will reflect both costs and buyer preferences.

The existing approach in Mexico is attractive as a strategy of arriving at a benefits package in balancing concerns for cost-effectiveness of the interventions covered (efficiency) with the need for providing financial protection against expensive treatments, some of which may not be as cost-effective. Specifically, the benefits package was broken down into three categories, and within a *specific category*, rare but expensive conditions, common conditions, and preventive action/health promotion—interventions that were the most cost-effective were included. Mexico has been able to cover conditions that account for 95 percent of all outpatient visits and most inpatient stays under its *Seguro Popular* insurance. Cosmetic surgery, transplants, and renal dialysis are excluded, but the latter is expected to be included in the package shortly. Mexico has a separate budget for prevention and promotion activity, including covering relevant portions of the *Opportunidades* program. Even here, cost-ineffective interventions such as rotavirus immunizations are excluded. And again, Mexico has proceeded in a step-by-step fashion, expanding the number of interventions as resources permitted.

Thailand too, has a similar approach. It offers a "comprehensive" package under its so-called *universal coverage scheme* which essentially covers "everything," except transplants and cosmetic surgery; and as in Mexico, it has a separate budget for health promotion and prevention, funded by an alcohol tax with revenues earmarked for this purpose.

India is probably not in a fiscal position to offer publicly subsidized benefits to the poor at a level as generous as that of Mexico or Thailand. This could mean restrictions on the types of hospital-based services that are offered. The limited inpatient care benefits offered under the RSBY scheme, ₹30,000 for a family of five annually, reflects this; as also the financial difficulties that Aarogyasri is currently facing with its more generous package of tertiary care. However, cost-effectiveness may still be relevant as a criterion, in defining which inpatient interventions could be supported by the scheme in question. Similarly, an upper bound for allocations for outpatient care coverage could be established, given the specific concerns about moral hazard, and moreover, cost-effectiveness considerations could be introduced.

Targeting the Poor

If resource constraints lead policymakers to direct health subsidies toward the less well-off, whether in a single or multiple pool system, the need for targeting is obvious. Various methods have been used for this purpose, including the use of means (or proxy means) for identifying the poor as has been done in some states under the RSBY program and in Mexico in the *Seguro Popular* program to decide upon premium contributions of enrollees; identification based on geographic location as in the case of Vietnam for the purpose of user fee exemptions and social health insurance; and targeting with the help of local communities, e.g., at the village level. Another possibility is self-identification by means of a "separating equilibrium" where only the poor are likely to come for the benefits, an approach followed by the Mahatma Gandhi National Rural Employment Guarantee Scheme. These methods are often deemed necessary because of the large numbers of informal sector workers whose earnings are hard to track and the generally poor compliance with income tax laws in India.

Krishna (2007) assesses these identification methods, all of which, he suggests, have a high likelihood of type I (exclusion) and type II (inclusion) errors, apart from the cost of collecting and maintaining the information "current" over time, given the dynamic nature of poverty. Rajasekhar et al. (2011) found serious problems with the identification of the poor in the rollout of the RSBY scheme in Karnataka recently which relies on a proxy means method to identify the poor. In the presence of means tests, if the government subsidizes the contributions of the poor (or if there are separate government programs for the poor), better-off informal sector workers, whose earnings are not easily tracked, would have an incentive to mimic the economic circumstances of their less well-off counterparts.

This calls for blunter targeting strategies. In Thailand, the 70 percent of its population not employed in the public sector is covered by the UHC scheme for the less well-off, with all premiums paid from general revenues. Another approach was taken by the Aarogyasri scheme in Andhra Pradesh, where anyone holding a ration card was defined to be poor. This liberal definition led to more than 80 percent of the Andhra Pradesh population (65 million people) becoming eligible to enroll in the program. Alternatively, a relaxed proxy means test could be devised where the criteria of no home ownership, no car ownership, or the lack of other difficult-to-hide assets could be used to define groups targeted for public subsidies. For example, the National Council for Applied Economic Research (NCAER) estimates that about 10 percent of Indian households own cars, and even if we include ownership of two-wheelers, the total of two categories about 40 percent of all households, which could lead to about 50–60 percent households being classified as poor (Desai et al., 2010). Use of such asset-ownership indicators for which ongoing administrative records are more likely to be maintained and updated, would also help reduce inclusion errors arising from households remaining on the rolls of BPL groups, well after they may have improved their economic situation.

How Should the Benefits Be Financed?

Previously we examined the question of who ought to pay for their benefits and who should receive a subsidy, concluding that in a resource-constrained environment, public subsidies ought to be directed toward the poor, except for health services involving significant externalities, or interventions that may be underutilized because individuals are unable to adequately account for future health consequences of their actions (e.g., health checks). Here we look at the specific mechanism of financing—such as general revenues, earmarked payroll taxes, household prepaid contributions for insurance, earmarked taxes on alcohol and tobacco, and other options.

The three major sources of financing are general revenues (earmarked) payroll taxes and out-of-pocket payments for health services. The most common way in which general revenues finance health services in India and in other Asian countries is via allocations to public sector health facilities. Payroll taxes are used to contribute premiums to social health insurance and found in most Asian countries, including in India, under the Employees State Insurance Scheme (ESIS). Some countries also rely on prepayment into voluntary insurance schemes. Table 1, based on the most recent National Health Accounts for India, shows that the share of out-of-pocket spending in total health spending in India continues to be high (with the remainder comprising government spending, private insurance, and external aid). India's out-of-pocket spending share is certainly much greater than some of its natural comparators, such as Brazil, China, and Indonesia which have relied on a combination of general revenues, payroll taxes, and social insurance.

TABLE 1. Share of Private Sector and Out-of-pocket Spending on Health (2005)

Countries	*Private expenditure as percentage of total health expenditure*	*Out-of-pocket spending as percentage of total health expenditure*
Bangladesh	70.9	62.6
Brazil	55.9	30.5
China	61.2	52.2
India	78.1	71.1
Indonesia	53.4	35.5

Source: Mahal (2010), from WHO (2008) and Government of India (2009).

Note: Private expenditure includes both out of pocket spending as well as contributions from private enterprises, for profit and not-for-profit.

Given that out-of-pocket payments above a certain threshold are undesirable for reasons of financial protection and inefficiency in resource use, general revenues and payroll taxes potentially constitute the main sources of financing health services for expanded coverage, with some limited potential for other kinds of earmarked taxes (on alcohol, tobacco, etc.). With the limited size of the formal sector (more than 90 percent of the employment is in the informal sector), "solidarity" contributions from payroll tax revenues of formal sector workers to financing health care for the poor are not a viable option at present, unlike in Colombia where a small share of funds from the contribution-based pool for formal sector workers has been transferred to support the pool that finances care for the less well-off. Nor would it be easy, given serious income reporting problems, to rely on earmarked income tax revenues from the broader group of "nonpoor" (presumably extending beyond the formal sector, possibly excluding low-paid formal sector workers) to contribute to a pool for the poor. Thus, general revenues are the most appropriate strategy to finance (insurance coverage for) hospital and outpatient services for the poor in India. Examples of this type of financing for health services include the Aarogyasri, Yeshasvini, and RSBY schemes in India, the insurance scheme for the poor in Indonesia, and PhilHealth in the Philippines. Hsiao (2008) also highlights the point made earlier that general revenues financing, if based on progressive taxes, can result in cross-subsidization from the rich to the less well-off.

For the nonpoor, two options are possible for curative care coverage. The first is to rely on payroll taxes for formal sector workers and (voluntary) prepayments by "nonpoor" informal sector workers to fund a separate pool for the nonpoor without any (or limited) public subsidies. In some versions of this option, the "nonpoor" in the informal sector are left out if there already exist mandated insurance schemes for formal sector workers—as ESIS in India, and similar schemes in Indonesia and elsewhere. For the nonpoor informal sector workers then, the options could be participation in voluntary prepaid insurance provided by private insurers, regulated so that community risk rating is used for setting premiums, as in Australia. Mechanisms to address adverse selection will be needed here. In Australia, premiums that increase by 2 percent above the standard for each year that individuals delay purchasing insurance after age 30, and any exit from being insured results in the individual as being treated as having no prior insurance, should they seek cover later. The second option is to have the "nonpoor" (formal sector and others), simply pay an unsubsidized premium to access the same health services that the poor are getting at subsidized rates. To the extent that participation by some of the nonpoor may be voluntary, safeguards will be needed to avoid adverse selection. Equivalently, one could imagine the poor enrolling in an existing scheme for people in the formal sector, except that their premiums are paid for by the government from general revenues (one example is the Republic of Korea).

The precise model adopted would depend on whether fiscal resources enable both the poor and the nonpoor to enjoy the same benefits package. Single pools face the risk of deadweight losses if differential premiums result in strategic behavior related to labor market participation (formal versus informal sector participation). Differential benefits in multiple pools may also be necessary to ensure political acceptance, or for historical reasons. Because contributory schemes for the urban-based formal sector employees have existed for a length of time, countries aiming to achieve expanded insurance coverage for the rest of the population tend to start out with separate pools for the informal sector (and the rural population), such as China under its New Cooperative Medical Scheme, and the examples of Mexico and Thailand discussed earlier. Efforts to combine the different pools in Thailand have faced significant resistance, and it looks unlikely that any move will be made in the near future to bring them together (Piya Hanvoravongchai, Chulalongkorn University, personal communication). Thus, we come down on the side of separate pools for the nonpoor informal sector population and formal sector workers as a matter of feasibility. It is sometimes suggested that potential gains in terms of economies of scale from a single pool can justify relying on a single pool, but the recent literature on this subject shows that these economies can be enjoyed even without such a step (as discussed later).

Funds for prevention and health promotion activities could come from general revenues given the significant differences in the perceived private benefit and the social benefits associated with them. In principle, one could rely on subsidies for the poor and regulatory direction for the nonpoor for some of the preventive activities, but it is unclear how effective such regulation might be in India's environment.

Some authors have proposed using revenues earmarked taxes on alcohol, tobacco, and fast-food as an additional source of financing (e.g., Reddy et al., 2011), whereas others do not support it (HLEG, 2011). Thailand funds some of its spending on prevention and health promotion programs from these taxes, but there are questions on their effectiveness in the Indian context. The federal structure of the Indian state, together with the fact that alcohol taxes are one of the few revenue sources over which state governments have some control, suggests earmarking revenues from these taxes could be a contentious issue, unless appropriate center-state compensatory mechanisms are designed. Resistance can also be expected from the tobacco growers and the fast-food industry for tax proposals directed at their businesses. Crucially, revenues from these taxes are sensitive to the price elasticity of demand for "sin" consumption, which has been shown to be quite large in India, and this has implications for the reliability of this source of revenues for health services (Mahal, 2000). That said, taxes on alcohol and tobacco could still serve as a useful tool for curtailing their consumption.

Managing Pooled Funds

A number of different models exist in the region. In India, an autonomous social insurance authority oversees the contribution-based Employees'

State Insurance Scheme. The RSBY scheme is managed by private insurers in states, whereas Aarogyasri health insurance is managed by a quasi-government organization, Aarogyasri Health Care trust. In some countries, Ministries of Labor manage insurance pools, and in Indonesia, a for-profit agency manages multiple schemes: contributory schemes intended for civil servants and the informal sector, as well as noncontributory schemes funded from general revenues for the organizations of local governments. Finally, ministries of health manage budgets that are allocated to public health facilities, or to services purchased from the private sector.

In thinking about the appropriate entity to manage funds, the main objective has to be the efficient management of funds, which requires attention to a number of tasks, especially funding the provision of health services to individual members in need, purchasing of health-care services, and maintaining quality of health services funded from the pool. Correspondingly, there is a need to avoid political interference with what is essentially a technical activity. The RSBY also requires its insurer-managers to enroll beneficiaries into the program, whereas Aarogyasri currently does not require any separate enrolment apart from having a valid "below poverty line" card. Another objective is that any surplus funds should be invested in a way to earn returns without too much risk. This could be important in cases where premiums are set on a life-time basis. (Germany is a prominent example.)

Underpinning the above activities is the ability to organize enrolment, design payment plans, to undertake contracting and legal support activities, designing and overseeing information systems that are appropriate for tracking costs of services, individual providers, beneficiaries, medical records, and for undertaking any other monitoring and evaluating activities, including research capacity. In India, there is now considerable experience in both public and private insurance companies with insurer-managers in Aarogyasri, RSBY, and the ESIS, and in general, so that some of the required skill sets are available, including in accreditation and regulatory roles. Additional support comes from the work of the Unique Identification Authority of India (UIDAI) that should also help in tracking enrollees more effectively and limit fraud.

We believe that one key area where further expertise is needed relates to the ability to design payments for health services in ways that serve the objectives of good quality, especially performance-linked payment systems. As of now, RSBY and Aarogyasri have followed rather blunt strategies of "blacklisting" noncompliant providers for fraud, or cases of blatant deficiency of service. Provider payment is a technical subject in its own right and is addressed in the next subsection.

Also crucial is ensuring that fund managers themselves have the necessary incentives to carry out their responsibilities effectively. This is not always guaranteed as suggested by low enrollment and claims rates observed by Rajasekhar et al. (2011) in the process of the RSBY rollout in Karnataka by private insurers. There has also been some concern in the context of RSBY and Aarogyasri that short-term profits were a major attraction for participating private manager-insurers. Specifically, given the per capita basis (per enrollee) on which the government contributes to these schemes has the potential of generating high short-term profits for companies owing to the low claim rates as enrollees are not always aware of the health services covered by them, at least in the initial stages following their launching. Some of these problems led to the takeover of management responsibilities by government-controlled Aarogyasri Trust in place of a private insurer-manager. However, it is not obvious that increased accountability and better performance will automatically result from this change. Increased accountability requires some form of competition for the "right to manage," as is currently happening under the RSBY scheme, and it may require additional oversight by a mix of the public sector, civil society, consumer representatives, and other stakeholders. We would argue against direct management by a government entity, given our concerns about frequent government interference.

Ongoing efforts to provide expanded insurance coverage in India are likely to result in multiple pools. The emergence of multiple pools, in turn, will increase the risk of duplication of coverage across population groups and would limit the extent to which economies of scale and scope in the provision of such coverage can be exploited. Thus, one useful fund management model to think about in the Indian context is that of Indonesia, where multiple funds are managed under one entity with a single claims-processing procedure (for instance), although the funds themselves are specific to population subgroups. A recent report prepared at the request of the US state of Vermont that is aiming at a single payer insurance model, proposes just such a manager (Hsiao et al., 2011).

Paying Providers for Health Services

The World Health Report for 2010 suggests that significant savings may be possible, something on the order of between 20 percent and 40 percent of existing health spending, if health systems around the world were to function more "efficiently" (WHO, 2010). Achieving improved efficiency in resource use for health services of acceptable quality is obviously important for the credibility of a plan for expanded (or universal) coverage and contributes to program sustainability. It turns that how funds are used to pay for health services can have important implications for efficiency, while simultaneously impacting the quality of services provided.

An immediate benefit from large collections of funds is the potential bargaining power they have in purchasing health services, consumables, and equipment. Centralized drug purchases by the government of Tamil Nadu are often cited as a model for replication elsewhere in India, and the same argument could apply to the purchase of other arms-length transactions, including the purchase of health services (Das Gupta et al., 2009). However, it is possible to also structure the payments themselves in ways that could enhance efficiency.

The best known model of how providers are paid is fee for service (FFS) where payment is for each category of service they provide, which is also the typical mode of payment for health care in India. Alternatively, there are "prospective" payment systems, such as capitation systems, DRG-based payment systems, and salary-based and global budgeting. Under the capitation system, the provider is paid a fixed amount for every patient or enrollee, irrespective of the seriousness of the treatment. Payments can also be fixed on the DRG (diagnostically related groups) a patient is categorized in, specifically the disease category further subclassified by type of interventions and seriousness of condition. Examples of DRG-type systems can be found under Aarogyasri and RSBY. Salaries paid to medical personnel are also an example of prospective payment, and this is typical of Ministry of Health medical personnel. Finally, global budgets are a form of prospective payment found in many countries in Europe and in Taiwan that cap the amounts allocated to health sometimes nationally and regionally and sometimes on a facility basis (note this is common for public hospitals in India owing to their reliance on historical budgets). Usually, some form of DRG or capitation-based payment accompanies global budgets to influence provider behavior; and bonuses or other incentives are sometimes added to the basic payment structure, conditional on attaining specific performance targets (Shaw, 2008).

From a health-care provider's perspective, each patient is like a risky purchase, and for whom the cost of treatment is a priori unknown. Each payment mechanism affects the providers in terms of the financial risk associated with treatment, and this affects their response. FFS payment places the entire burden on the payer and none on the provider. Thus, FFS or similar systems are likely to result in over-treatment and little preventive activity. With no cost constraints to worry about, and as long as the provider values patient health, treatment would continue until the marginal benefit of an added intervention has reached zero. Thus, FFS is likely to result in high costs and treatment that is not cost-effective at the margin. The dominance of FFS as a mode of payment in India points to opportunities for large efficiency gains.

The reverse of the above is pure prospective systems such as capitation payment. Because the severity of the patients' condition has no effect on the payment, the incentive is for providers to choose the least risky (or least complicated) cases, to underprovide care, and to dump patients on other hospitals of the "last resort," usually public hospitals. Prevention activities are likely to be supported by providers under capitation-based systems because they lower the financial risk to health-care providers. While costs may be controlled, there is the risk of receiving low-quality care. DRG-based payment systems try to balance quality of care with cost containment by linking severity of health condition to payment. Even here, providers have an incentive to increase the number of episodes or bump the patient up to a DRG with a higher level of reimbursement, while undertreating patients within a DRG or episode category. "Mixed" payment systems involving some combination of DRG and actual cost-reimbursement have been suggested as ways to get around the problem of balancing quality and costs. Global budgets are useful in that they help in ensuring sustainability of the financing pool. When combined with further payment incentives to influence allocations across interventions, they can help in addressing allocation decisions. However, the concern about quality of services remains. Many government hospitals in developing countries run on a "historical cost" basis and, although such an arrangement (coupled with low budget allocations, which effectively mean global budget restrictions) ought to keep costs low, misallocation of resources can result as doctors have no incentive to allocate resources cost-effectively, or improve quality.

Developing country literature, although limited, provides some evidence that health-care providers respond to payment incentives. One study for China found that budgetary cuts to public health providers in the early 1980s led them to disproportionately increase services for which prices were not controlled by the government (such as drugs and diagnostics). The upshot was a rapid increase in spending on drugs and diagnostics and overall rapid health-care cost inflation (Wagstaff et al., 2009). Liu and Mills (2003) found that the introduction of a bonus system in Chinese hospitals and greater incentives for doctors led to increases in hospital revenues as well as treatments that were more expensive and capital intensive. A study for Thailand also found that FFS payment mechanisms were associated with rapid health-care expenditure increases (Shaw, 2008). Yip and Eggieston (2001, 2004), on the other hand, found that prospective payment systems in some Chinese hospitals led to a slower increase in health-care spending compared to other hospitals paid with an FFS system. High rates of absenteeism in many primary care facilities in India are potentially one reflection of a payment policy that emphasizes guaranteed salaries unaccompanied by performance incentives. Moreover, the large private sector presence in the provision of health services and their likely role in any health reorganization suggest that incentives will be a major driver of provider behavior in India. Thus, there is a case for India to rely less on FFS as a means of paying health-care providers on grounds of cost containment.

Evidence of the impact of payment mechanisms on quality is hard to find outside of developed countries, but is key for their viability. Patients cannot judge clinical quality on their own, although they may be able to better assess service quality, such as the availability of drugs, personnel, interactions with personnel, and infrastructure. Assessments of the number of personnel at different levels of expertise, the number of beds and specialties available, and other services can also serve as ex ante measures of quality of a health service facility (such as a hospital). Under the Aarogyasri scheme, a team of medical experts assesses the diagnosis before agreeing to the medical procedure for which reimbursement is requested by a health-care provider. However, ex post measures such as health outcomes following treatment are desirable as measures of clinical quality for obvious reasons, such as tracking patient health outcomes following discharge. In addition to heavy-handed instruments such as the blacklisting of health facilities as in RSBY and Aarogyasri, the use of "pay for performance" contracts—that include bonuses for meeting certain quantity and quality standards—may be necessary (Witter et al., 2012).

The choice of a specific payment mechanism is obviously conditional on how easily it can be implemented and this may require complementary policy steps. For providers to be able to respond to payment incentives, there must be a reasonable degree of flexibility in how they provide their services in response to the mode of payment, which is not an issue for private and nongovernmental providers, but is for public providers which will require greater autonomy of operation, especially in terms of "residual" claims over resources. This may necessitate additional reform in the public sector, over and above recent changes that have allowed public hospitals to retain some of their reimbursements from RSBY and Aarogyasri, in addition to Ministry of Health (MOH) budgetary allocations. Information on the cost of service provision will be needed to guide funders in setting reimbursement rates. If reimbursements rates are lower than cost, providers may not agree to work with the financing agent, or may resort to "extra-billing." If reimbursement rates allow for large variations in net margins across different services, incentives will be created for providers to disproportionately provide the more profitable services, and allocative inefficiency may result. Data on outcome measures will be necessary to set up performance-based payment systems. Thus, some payment systems can be extremely resource intensive (including needing human capital) to implement. Shaw (2008) suggests that salary-based systems and capitation fees are relatively easy to implement. However, payments made on an FFS basis or DRG-type systems put enormous informational burden on payers and health-care providers owing to the complexity of treatments, emergence of new treatments and technology on an ongoing basis, the need for assessing the effectiveness of new technologies and so forth. Pay for performance systems are difficult to administer, because they require a monitoring and measurement capacity related to (complex) outcomes. Finally, the state of the Indian legal system that contributes to poor enforcement of contracts may also need addressing if performance-based contracts are to work.

A final issue is whether provider payment methods should vary by category of service. Some type of prospective payment system is desirable for inpatient care as a mechanism to control costs of (complex) care, while allowing for adjustments associated with the complexity of a case as a mechanism to address quality, given how expensive it is. Outpatient services raise a number of tricky issues: such as balancing the need for providing necessary services with containing moral hazard on both the demand and supply sides, serving as a screening device for referral to hospitals (gate-keeping), and providing preventive care. Capitation on a per enrollee basis is attractive, possibly with some adjustment for "case-mix" indicating the likely riskiness of the patient (age or sex) and funds following whomever the enrollee chooses as the primary provider. However, there is the risk that primary care providers will try to offload patients to higher-level facilities. Without appropriate safeguards, the capitation system will resemble the state of referral systems in India where the primary health system has effectively abandoned its gate-keeping role. For the capitation payment model to work, a more "integrated" provision of outpatient and inpatient care is required and could be encouraged by making reimbursement from pooled funds conditional on having an integrated provider network. However, there remain at least two challenges to the workability of the capitation system for outpatient care in India: the lack of access to qualified primary care providers in rural and remote areas and the informational requirements for tracking health service quality. To balance the goal of maximizing public health with efficiency considerations, it would be useful to have FFS payments for preventive visits (e.g., for immunizations, health check-ups). The remaining categories of outpatient care could be controlled with a designated upper limit on spending for each household.

Paying the Consumer of Health Services

We have not focused on payments to the consumer thus far. In fact, there are reasons to consider paying the consumer to use specific types of services in some circumstances. These include providing incentives to individuals for undertaking preventive visits and for health promotion activities to cover any opportunity costs associated with such visits. Cash transfers provided *conditionally* to parents (mothers) as incentives for obtaining immunizations for their children and maternal and child health services, as in the *Opportunidades* program in Mexico, or the Janani Suraksha Yojana (JSY) in India, are examples of consumer payments of this kind. There is some evidence that JSY has led to an increase in the use of maternal health services in India (Fan et al., 2011). As another example, Australia awards $125 to households for completing child immunizations. In principle, the same argument could apply to interventions directed at adult behaviors that have long-term implications for their health, such as food consumption habits and physical activity, and for health check-ups, at least among the poorer members of the population.

Some authors have suggested that direct (unconditional) cash transfers including pensions be provided to poor households, which could be used for any purpose, including outpatient health care and nutritional services (e.g., Duflo, 2003; Panagariya, 2008). The main advantages of cash transfers are: administrative hassles and monitoring costs are minimized (as opposed to vouchers) and deadweight losses may be lower. They could also increase the accountability of providers to patients given the purchasing power is in the hands of the household and allow consumer flexibility in choosing providers especially when (funder approved) providers are not readily available. Both demand- and supply-side moral hazards are limited by the fixed amount of the transfer (Panagariya 2008).

The flip side is that unconditional transfers are unlikely to influence behaviors with externalities. Moreover, if individuals are not good judges of their own health, the effect of such transfers on health care–seeking behavior will be less than its desired levels. Relying on households to choose their own care also means foregoing economies of scale that could be exploited by bulk purchases by someone on their behalf. There is also the risk of gender bias in health-seeking behavior when transfers are unconditional. And, the use of vouchers/conditional transfers would also curb demand-and-supply-side moral hazard if accompanied by an upper limit to the voucher amount (with the proviso that in a corrupt environment, some siphoning off of resources by the administrative machinery is always possible).

In general, we support the idea of transfers conditional on preventive visits to health-care providers to unconditional transfers. The only study that we are aware of that compares the two approaches (and does so in a randomized evaluation framework) is by Akresh et al. (2012) for Burkino Faso, finding that conditional cash transfers had a significantly higher effect on preventive care visits than unconditional transfers and, moreover, that the latter had no effect on the number of preventive visits. That brings us to outpatient health services that do not have a strong prevention focus. There is very little evidence, one way or the other, of the effect of unconditional transfers on health-seeking behaviors relative to conditional funding, despite their many attractive features. We propose that pilot evaluation studies be carried out to investigate this issue further.

Supply Side Issues

Much of the preceding discussion has focused on the demand side of the health sector, with only limited attention paid to supply, mainly in the context of concerns about service quality related to provider payment systems. There are at least *three* other supply side issues that need careful attention in the context of policies aiming to expand coverage in India which we alluded to previously. The first has to do with redefining the role of public sector health services in India, given significant private sector presence and a need for payment mechanisms that can elicit improved health service performance. The second has to do with ensuring the supply of health services (including personnel) of adequate quality to meet the demands of population, particularly in rural and remote areas. The third has to do with regulatory capacity and oversight mechanisms. We will focus primarily on the first two, since the third has been examined in some depth in Reddy et al. (2011) and the HLEG report.

Public Health-care Services

The HLEG report for India emphasizes public sector health-care providers as a central plank in its proposed model of universal coverage, and the role of the private sector will essentially be to "complement" public providers. In this vision, as we see it, funding from public pools could be used to pay for private care that fills gaps in public provision (HLEG, 2011). The proposals in HLEG report do not suggest a high level of autonomy for the public sector, although they do seem to go beyond what exists at present. It is well known that limited autonomy and community oversight along the lines of the existing Rogi Kalyan Samitis (and equivalents for lower level health facilities) where public providers possess flexibility with regard to the use of any revenues they raise on their own (say via reimbursements from insurance or user fees) has not been effective (Gill, 2009). In the HLEG report, increased accountability of government facilities is to be ensured by a cautiously phrased "performance-based management" and a greater community and local government role in oversight of health facilities.

We do not believe that the proposed HLEG model is enough, given the public sector performance on record. To be sure, there are examples of acceptable public sector hospital performance such as Kerala's public hospitals that seem to have operated effectively in the presence of the private sector, including competing for RSBY funds, but Kerala is likely an outlier with a highly informed population, and a long history of decentralized activity and public sector performance there has not been easily replicated elsewhere in India. We believe that public sector responsiveness to competition from the private sector is another pillar in influencing the former's performance and the current approach does not consider this. Indeed, even in Kerala, the decks are stacked against the private sector because public facilities have separate budget allowances which effectively subsidize their provision of health services and adversely affect the economic attraction of RSBY reimbursements to the private sector, similar to the situation in Thailand. In Thailand, most hospitals are owned by its Ministry of Public Health (MOPH), and while salaries are covered by the Ministry, funds for nonsalary expenditures are based on a mix of revenues from contractual arrangements with risk pools and user fees (Hawkins et al., 2009).

A more pertinent observation is that the private health sector in Thailand is quite small relative to its public sector, a situation very different from India's. It is difficult to imagine that the private sector in India can simply be left out of a major role in the provision of health care services under plausible scenarios (Banerjee et al., 2008). The very emergence of the private sector has been in large part due to a nonperforming public sector, and it can be imagined that without a competing (and not a complementing) private sector, the public sector would have even less incentive to improve performance. In this context, recent changes in the Aarogyasri scheme in

Andhra Pradesh that have limited specific types of covered care only to public providers reflect the strategy outlined in the HLEG report, and it will be of interest to track the associated implications for quality of health services in public hospitals.

If public and private providers compete for reimbursements from pooled funds, and this is the likely future scenario in India, the preceding discussion on provider payment mechanisms raises two questions of interest. If a large fund is willing to contract with both public and private providers, how can public providers compete effectively with private providers?

We do not think that Indian public facilities can serve as credible competitors to private providers in India, at least in their current organizational form and accountability mechanisms. Instead, considerable health facility autonomy in terms of resource allocations is warranted. Payment systems can be used to influence health facility performance much more directly to achieve goals of equity and cost containment in arms-length contracts with autonomous facilities than where managers and other medical staff are paid government salaries on a permanent basis. And autonomous facilities have greater motivation to use their resources efficiently if they are the residual resource claimant. For example, reforms initiating a purchaser-provider split were introduced into the NHS in the United Kingdom in the early 1990s, and as part of the process many public hospitals became "trusts" with a degree of control over staff hiring, services offered, and right to retain financial surpluses (McPake and Hanson, 2004). For their income, they were dependent on contracts with district health authorities, General Practitioner (GP) fundholders, and private patients. Nontrust hospitals did not have this freedom. Studies found that productivity increased in trust hospitals relative to nontrust hospitals. Over the last decade, Poland has converted several hundred public hospitals into autonomous status with a high degree of decision-making and surplus retention authority (Schneider, 2010). Revenues for these hospitals were obtained from DRG-based payments (for inpatient care) and capitation (for outpatient) from insurance funds; and state budgets covered certain types of highly specialized care, emergency care, blood banks, and capital costs. Public hospitals that have become autonomous in Poland have seen surpluses and staff salaries go up. Again, available evidence points to increased productivity. Information on quality impacts has proved more difficult to find for both the NHS and Polish reforms.

Another relevant experience is that of China, where public sector facilities rely more heavily on user fees than their Indian counterparts, which has effectively turned them into private providers. There the government moved to providing only a small part of their recurrent expenses following the introduction of market reforms in the early 1980s, with the rest expected to be made up via charges to patients or their insurers. The consequence has been supplier-induced demand for diagnostics, branded drugs, and other services that did not have price controls and that could yield greater surpluses, at the cost of services for which there were price restrictions imposed by the government (Yip and Mahal, 2008). The consequence was resource misallocation and health expenditure inflation. Studies in China have also shown that the introduction of suitable provider payment mechanisms can ameliorate these effects and this means that public facility autonomy and payment systems have to be in sync (Yip and Eggleston, 2001, 2004). China's experience also underlines the need for oversight bodies to direct autonomous public hospital functioning in line with social objectives.

Indian public sector health service providers are likely to be faced with cost inflation pressures of their own. There is the risk of losing top medical professionals in public health facilities to the private sector and its high salaries, without raising earnings opportunities within the public sector itself. Already this is happening in Malaysia which has experienced a significant drain of its top specialists to the private sector (Mahal, forthcoming). Moreover, public sector health facilities would have to keep pace with technological developments in order to stay competitive with the private sector. Certainly, in the case of Colombia whose health reform in the 1990s included a vision for public sector competing with the private sector has found a public sector that has become a less effective competitor over time. In the absence of significant operational and funding autonomy, public sector facilities in India are likely to see a slow death. But granting autonomy to public sector health facilities has not been politically easy when employees have guaranteed jobs, as indicated by previous efforts at increased autonomy of public hospitals in the state of Andhra Pradesh in India.

Finally, we believe that the HLEG recommendations on introducing professional cadres related to public health and managerial functions in the public sector are sound. They fit well with our ideas for an autonomously functioning public sector. But its proposals on autonomy need to go further if the public sector provision is to be envisioned as playing a significant role in the health over the longer run. What types of models of autonomous public facilities are appropriate for India? While we are not firmly wedded to any one set of proposals, there are at least two possible candidates. One model is some form of "integrated" group of primary providers and first-level hospitals (say, at a district level), similar to Thailand, perhaps competing with private sector network(s). Alternately, the model could be of autonomous networks of primary providers competing with private sector networks, both linked to separate autonomous first-level and tertiary public hospitals (or private hospitals) via a referral chain. This is the type of model being proposed in the context of Malaysian health reforms.

Another critical set of issues has to do with ensuring "fair" competition between the public and private sectors, given the former may have added social goals, and that subsidies from the government may be reaching public hospitals. Policymakers will have to tackle questions relating to subsidies related to the wage bill and investments in infrastructure and investment that tend to go to public facilities and, at the same time, take account of the fact that public hospitals sometimes have unique responsibilities, including serving as providers of the last resort, and often treating more complicated cases. This is why payment systems that reflect patient risks (such as DRG mechanisms) are such a critical element.

The major bottleneck, as we see it, is not in India's urban areas, where there already exists a high concentration of both public and private providers. It is in rural areas where shortages of both groups of providers exist, and this can potentially create problems for the type of competition we have in mind and associated accountability for performance. Here, alternative models may be appropriate, and these might allow for both private and public providers to deliver services collaboratively.

Health Services in Remote and Rural Areas

A major challenge in expanding social health protections to rural and remote areas in India is the lack of availability of providers there. This is well documented for the public sector. According to the Ministry of Health and Family Welfare (MOHFW) publication on rural health statistics, as of 2010, there was a shortage of 2,115 rural hospitals (Community Health Centers) or about 33 percent less than required as per its own norms, and a 15 percent shortage from an estimated required level of 158,000 (MOHFW, 2011). The data also indicates shortages of between 50 and 70 percent of physicians and various specialists, lab technicians, and radiographers at community health centers. Amenities such as water supply and electricity were reported missing at between 10 and 15 percent at primary health centers. This is the situation after five years of the NRHM which involved substantial infusion of new resources into health services in rural areas. In addition, there are well-known problems with absenteeism in public sector health facilities (Chaudhury et al., 2006). The option of the private sector is not without its limitations either, when it comes to improving accessibility of health services in remote and rural locations. The quality of private sector services available to the population living in these areas also appears to be poor, with many residents, especially poorer ones, relying on unqualified providers.

Under the NRHM, there have been efforts to improve public sector provision in underserved areas by hiring locally, providing local hires with basic training, and adding incentives to enable them to supply some basic treatment and referral services. Budgetary allocations have also been increased to public facilities, along with some added autonomy of functioning and community oversight. It is too soon to say whether this approach is working, although at least one recent evaluation is somewhat pessimistic on this front (Gill, 2009). An alternative strategy has been tried under the RSBY and Aarogyasri schemes, namely, requiring empanelled private hospitals to provide mobile clinics and conduct health camps, especially in underserved areas, as part of their contractual arrangements. However, it appears that the camps mainly serve as a route to refer patients to private hospitals (International Labour Organization, 2011).

The HLEG report suggests other steps: better management of human resources in the public sector via the development of a public health cadre that would take the pressure on existing clinical staff for activities related to management, prevention and health promotion activities; training and hiring more staff including the introduction of three-year degree programs to train medical staff equipped to work in rural areas; and setting up more nursing colleges and medical colleges in less served areas. These are all good ideas, especially since it is well known that current (highly trained) medical graduates do not always want to be located in remote areas, and the location of the training institution tends to influence where its graduates ultimately locate (Lagarde and Blaauw, 2009). Other approaches have been adopted by Sri Lanka and Vietnam that allow for private practice by public doctors outside of work hours as a way of making rural locations attractive. Thailand, Malaysia, and Sri Lanka require their new medical graduates to serve in rural areas following graduation and have experienced relatively high compliance, at least for graduates of highly subsidized medical colleges. However, countries where private medical schools dominate are likely to have less success in imposing such service requirements, and this is India's situation.

The recommendations of the HLEG report with respect to increasing human resources for health in rural and remote areas—formation of a new class of rural health practitioners, adding more staff, opening training institutions in poorly served areas—are also pertinent. However, they could take a long time to implement, and require investment of significant amounts of new resources in training infrastructure. With the exception of the focus on local training institutes and the development of a cadre to take on responsibilities from clinical staff, the HLEG report did not focus on incentives that influence public sector personnel location behavior and performance to the extent that it should; and it tends to underplay the potential importance of the private sector in increasing coverage. It is true that private providers generally find service in rural areas unattractive, but this is not the case with the large body of unqualified workers who account for a considerable share of health-care provision there, and could number well in excess of 1 million in the country (Panagariya, 2008). Of course, similar ideas are applicable to lower level staff placed at public facilities in rural areas, for whom additional training to provide some types of simple curative care may be appropriate; in addition, their earnings could be tweaked to allow for performance incentives.

We think there may be scope for alternative approaches in the short run, even if we adopt the HLEG recommendations as the model for the longer run. Berman (1998) mooted the idea of formally bringing in "less than fully qualified" private providers into providing services in a more regulated manner. This is a desirable approach because less than fully qualified providers will continue to be attractive as a source of health care for millions of people, until more accessible, affordable, and higher quality alternatives become available, and they are also the individuals who have a track record of providing services in rural areas. They could potentially contribute to delivering immunization programs or health promotion programs, on some form of FFS or per capita payment system. In addition, we believe that this group could be a potentially rich recruitment base for short-duration and three-year training programs of the type envisaged in the HLEG report. Following such training, they could be engaged in a broader range of health-care services financed from various pools.

Some authors have suggested the "strategic use" of financial incentives to influence the location of medical personnel, although little evidence on this subject is currently available in developing countries. Bärnighausen and Bloom (2009) suggest using scholarships and loans for medical education that are *conditional* on the individual working in a remote/rural area upon graduation for a specific number of years. They present evidence, primarily for developed countries, suggesting that program participants compared to nonparticipants were more likely to work in areas that were designated as "underserved." Other alternatives that do not involve the location of highly trained medical personnel in remote areas rely on solutions based on an information technology platform, sometimes called "telemedicine." These include phone-based medical consultation services such as those provided by the Health Management Research Institute (HMRI) in Andhra Pradesh in India. An approach now being piloted in Africa and India is the use of treatment protocols on hand-held devices by locally based individuals. The treatment protocols can readily be tailored to act as a referral tool for more complex cases. One example of such an effort is that of CARE, a major operator of hospitals in India, which is piloting a scheme using hand-held treatment protocols in the 50 least developed villages in Yevatmal district, one of the poorest districts in the western Indian state of Maharashtra. Similar pilot programs that involve some payment for services rendered by individuals carrying these devices are being tested as well in a number of countries. Yet, in any case, the incentive structures in place for these new technologies need to be considered, as new technology alone is not a silver bullet.

Financial Implications of Expanding Health Coverage

Expanding coverage to large numbers of people is not just a matter of rights or social well-being. It is also a matter of what a country can afford financially as reflected in trade-offs with other options, including nonhealth investments. Indeed, resource constraints are acknowledged in the human rights literature as implying the "progressive realization" of rights, or more precisely, restricting interventions only to *Pareto-improving* options. That is, the rights literature takes an even more rigid view toward trade-offs than of most economists. While discussions around expanding coverage in China, India, and elsewhere in Asia reflect the region's rapid economic advancement and associated fiscal prowess, there are other compelling priorities that compete with the health sector for resources. India's recent rapid growth of tax revenues has underpinned an expansion of its governments' resource "envelope," yet even within the so-called social sector, the government must contend with competing demands, such as the Mahatma Gandhi National Rural Employment Guarantee Scheme (MGNREGS) and the implementation of the "Right to Food." In the year 2010–11, support for the MGNREGS exceeded ₹40,000 crores; and estimated public subsidies for food could amount to as much as ₹90,000 crores, which is roughly India's current public spending on health (Government of India, 2011b). An assessment of the financial requirements of expanded coverage of health services for India, and especially its implications for public and private financing is relevant from this perspective.

Multiple estimates of spending requirements for expanded levels of health coverage are currently available for India, although the methodology is not always clarified and calculations tend to be incomplete. We assess these different estimates and seek to arrive at what we consider a reasonable accurate picture of what the financial implications are, both in terms of spending required, as well as the share the government will need to contribute.

The HLEG report assessed the fiscal implications of "progressively moving" to universal coverage in India and concluded that adopting such a strategy would require public spending on health of 2.4 percent of GDP by 2017 and 3 percent by 2022. This amounted to ₹675 per capita in 2011–12, rising steadily to ₹3,450 by 2021–22 (all at 2009–10 prices) (HLEG, 2011, p. 106). The report itself, however, did not provide enough information to determine the basis for these numbers. To be sure, there are projected trends in additional training costs; infrastructure and operating costs reported in various sections of the HLEG report which, if combined with existing spending on health by the governments at the central and state levels, could explain its projected share of GDP devoted to health spending, but the relationship between the different projections is tenuous.

An earlier study, the report of the National Commission on Macroeconomics and Health (NCMH) (Government of India, 2005) also sought estimate public spending for expanded population coverage. The NCMH report does a better job than the HLEG of sourcing its estimates, even though there are discrepancies at various places. Annexes I–X and Table 4.2 of the NCMH report suggest that the expenditures required to achieve a minimum level of coverage at the primary and secondary levels of care would have been roughly of the order of ₹103,350 crores in 2005 (current prices) (Government of India, 2005, Annexes I and II, pp. 135–39). The coverage in question amounted to ensuring that a specific set of health conditions would be fully provided for at health subcenters, primary health centers, community health centers, and district hospitals. The aggregate cost of this coverage was calculated by multiplying the cost of standard treatment protocols for specific covered conditions by the total number of estimated cases of each condition in India. However, it was also assumed that only 70 percent of the actual cases would present themselves at this aggregate cost—in effect, assessing that health services would be inefficiently provided. That is, actual unit costs would be roughly 40 percent higher than estimates based on the cost of a standard treatment protocol. Overall, estimated aggregate costs of expanded coverage amounted to about 3.1 percent of India's GDP at factor cost in 2005–06, which is roughly what the HLEG report projects at the time universal coverage is achieved in 2022 (HLEG, 2011; Government of India, 2011a).

Both sets of estimates reported above have limitations. First, as the NCMH report clearly acknowledges, its unit cost estimates were based on staff being paid government salaries. If private sector earnings data was used to assess the cost of human resources to address incentive issues peculiar to the public sector, the requisite expenditures could be as much as 30 percent higher, or closer to 4 percent of GDP, according to the same report. Second, both the HLEG and NCMH reports did not consolidate expenditures on health promotion, research and development (R&D), and regulatory oversight with the costs of treatment when calculating required levels of spending for expanded coverage. Third, expenditure estimation exercises in both the NCMH and HLEG reports do not explicitly account for changes in the epidemiological profile or model of care over time. This would effectively rule out the impacts of health transitions, changes in utilization of prevention interventions, or rising demand (and supply) of high-end care, the last being a major driver of health spending in developed countries (Newhouse 1992). The reports do not directly link expenditure requirements for funding the NCMH benefits package to their recommendations for plan spending by the government. Both the NCMH and HLEG reports adopt the approach of estimating infrastructure and training expenditures independently of the benefits package, by defining human resources and capital requirements according to "norms" as stated by the government, implicitly assuming that fulfilling these requirements suffices for delivering the benefits package, without taking account of actual people needed to provide the package, issues of governance, and incentives for public provider behavior.

Prinja et al. (2012) provide an alternative set of expenditure estimates for universal coverage for India. The authors upwardly adjusted estimates of health services utilization from the NSS data in 2004 by almost 100 percent and allocated hospital outpatient visits and inpatient stays to different health conditions based on one year of data on outpatient visits and inpatient stays in four public hospitals located in a large city (Chandigarh). Standard treatment protocols were used to derive unit costs of treatment by type of condition and care received, but human resource costs were valued at private sector rates to take account of the fact that public sector salaries are compensated at levels much lower than would ordinarily be sustainable for incentive reasons. Moreover, Prinja et al. (2012) carried out sensitivity analyses to take account of interstate differences in morbidity profiles. Their estimates of the costs of outpatient and inpatient care at secondary and tertiary hospitals were then added to estimated expenses of primary care coverage in the NCMH report to obtain overall expenses for universal coverage ranging from ₹132,800 crores to ₹429,000 crores in 2011, or about 2.1 percent to 6.8 percent of GDP at factor cost, with a mean estimate of 3.8 percent (₹281,000 crores). As in the NCMH report, Prinja et al. (2012) did not assess the impact of technological change on health spending, or changes in epidemiological profiles over time. Nor did their study focus on starting and scaling-up costs, or spending on health promotion and regulatory oversight.

Consolidating the Findings

A midpoint estimate of the annual operating costs of a program that provides expanded coverage of health services at the level of primary, secondary, and tertiary care would be around 4.3 percent of GDP (at factor cost) based on available estimates, based on findings from Prinja et al. (2012) and the NCMH report. This is because to be consistent with the NCMH report, Prinja et al. (2012) should have adjusted the cost of primary care for which data was obtained from the NCMH report upward by 90 percent, since NCMH valued human resources at government staff salary rates, and should have accounted for inflation, since the NCMH calculations were for the year 2005.

To the above estimate, we add annual operating costs directed to additional efforts at health promotion, R&D, regulatory oversight bodies; this was estimated to be ₹8,140 crores in the NCMH report for 2005 (Table 4.1 of Government of India, 2005). Assuming that the share of this component of spending remained unchanged (as a proportion of GDP), this would amount to an additional health spending of 0.2 percent of GDP in any given year, including health promotion expenditures of about 0.1 percent of GDP. Overall, therefore, "universal coverage" could cost roughly 4.5 percent of GDP, all else the same, or ₹330,000 crores in 2011.

In principle, infrastructure expenditures (including upgrading existing facilities) ought not be included in estimates of annual spending required to achieve universal coverage because the unit cost calculations in both Prinja et al. (2012) and the NCMH report already include an allowance for depreciation of equipment and buildings and estimates. One could make a similar argument for training costs if, as in the calculations of Prinja et al. (2012), human resources are valued in terms of earnings obtainable in the private sector. In this case, the earnings implicitly compensate for training costs. Both categories of additional spending (infrastructure and training) comprise significant additional start-up expenditures and will have to be paid for by some party, and this will usually have implications for the government budget.

Fiscal Implications

Ignoring start-up costs for the moment, the estimated total spending of 4.5 percent of GDP is unlikely to be financed solely by the government. About 26 percent of the aggregate expenditures comprise public goods (health promotion, R&D, regulatory oversight) as well as primary care which we can assume to be subsidized by the governments at the center and the states. As for the remainder, we have suggested earlier that subsidies ought to be directed only to the poorer groups, and to children. We adopt a generous definition of "poor" and assume that 50 percent of population is defined as poor to account for targeting tools that are blunt. Moreover, given that 31 percent of the population consists of children in the 0–15 age-group and some children are already included among the poor, curative care for an additional 15.5 percent of population would need to be subsidized from public funds. In sum, about two-thirds of the population could end up being covered by public subsidies for health services at the secondary and tertiary levels, including specialist outpatient services, and this translates into government health subsidies of about 75 percent of all projected spending for a program of universal coverage (3.4 percent of GDP at factor cost on an annual basis), including prevention, promotion, research, and regulation-related spending.

Start-up Costs and Scaling Up

The previous calculations refer to annual recurrent spending (including depreciation allowance) for fully fledged coverage once the program for expanded coverage reaches maturity. In practice, reaching this stage will take a number of years, especially considering the gaps on the supply side in the health sector, including setting up new infrastructure for provision of health services, the training of human resources, upgrading infrastructure for health services, research institutions, regulatory entities, and so forth. Unfortunately, none of the estimates that exist provide a reliable set of numbers to go with, owing to the inadequate linkage between the needs of the benefits package and the way start-up spending estimates were derived (essentially satisfying government norms). Nonetheless, projected expenditures in the HLEG report suggest quite substantial spending over a decade (starting 2011–12), averaging at about ₹9,500 crores annually for health services infrastructure and about ₹3,700 crores annually for training infrastructure over the same period.

Capital investments will need to be accompanied by additional operational expenditures, including for human resources, maintenance, utilities, consumables, etc. The speed at which the additions to infrastructure occur will drive associated operational spending. One estimate from the HLEG report suggests that over the period from 2012 to 2020, aggregate annual operational expenditures would need to increase to about ₹216,000 crores in 2020 from its level of ₹40,000 crores in 2012 (Figure 7 of HLEG, 2011). Although it is unclear whether the 2020 projections of operational costs are at current prices or constant prices, these lie well below the midpoint estimate based on Prinja et al. calculations for the scaled-up program, which would suggest operational expenditures of the order of ₹315,000 crores at 2011 prices (excluding health promotion, R&D, and regulatory oversight).

There remains the question as to who will bear the burden of these initial expenditures. It is possible, provided that staff earnings are based on social valuation/market rates where available, that the private sector finds it worthwhile to invest in training programs, and potential trainees would then find it worthwhile to pay. Of course, if staff earnings remain low, training programs will have to be subsidized by the government. The same applies to infrastructure development. Provided the government can set up credible guarantees for demand backed up by adequate returns, infrastructure investments could well be undertaken by the private sector health service providers, even in rural areas. If that is not feasible, the start-up expenses on infrastructure and training will have to be incurred by the government, especially in earlier years, where the focus will be on expanding facilities in underserved areas.

Comparisons of Expenditures to Mexico and Thailand

How valid are the above numbers based on what we know about other middle-income economies? Mexico and Thailand both recently embarked (in the last decade) on a program of greatly expanded coverage for their populations and could offer a guide to the expenditures involved. While health needs and consumer preferences are likely to differ across countries, both these comparators have significant numbers of poor populations living in rural areas, and both seem to have serious challenges associated with noncommunicable conditions, which is also the case in India. Both also rely heavily on government facilities to deliver the benefits package under their "universal coverage" plans and have good cost and budgetary statistics.

Each year in Thailand, the National Health Security Office (NHSO) produces cost estimates on the basis of which the government budgets its payments to health-care providers—estimates of per capita payment (adjusted for health risk) for outpatient care, estimates for per capita expenditures for inpatient stays, per capita payments for prevention and promotion, and other categories as a means to transfer funds to providers in province and district-level authorities (Suchonwanich, 2010). Salaries are paid separately and out of a central pool from the Ministry of Public Health of Thailand, amounting to about 55 percent of the total health budget in the most recent year for which data is available (Putthasri et al., 2011). Unfortunately, there is no information on costs of infrastructure development, but Thailand did invest heavily in public sector health-care infrastructure development prior to the introduction of the universal coverage scheme. A second set of estimates on per capita spending are available for Mexico, based on a specific list of health interventions, on a package of basic and medium-complexity interventions (involving both outpatient and inpatient services) and on high-end tertiary-level interventions of a more limited nature (also shown in Table 6 of Gonzalez-Pier et al., 2006). Although estimates of infrastructure and human resource requirements for Mexico are probably not very suitable for the Indian context, the data indicates that infrastructure expenditures increased by 150 percent between 2001 and 2006, the initial years of reform. Directly using the per capita numbers for the different types of health services in Mexico and Thailand (after adjusting for differences in income per capita) to the Indian population, we estimated universal cover to range between 2.2 percent and 4.5 percent of GDP at factor cost, with data from Thailand pointing to the upper end of the range of estimates, and Mexico to the lower end.

Prices, Incomes, and Technological Change

Although the estimates of health spending that we have used appear generous, the actual levels of operational spending required may be higher. The implementation of a universal coverage scheme would lead to a significant reduction in the price at which individuals are able to obtain health-care services. This led Prinja et al. (2012) to assume that health services utilization would be roughly double the rate reported in the 2004 NSS data.

Under the RSBY insurance program, which has a fairly limited cover of ₹30,000 for a poor family of five, beneficiary hospitalization rates were of the order of 2.7 percent. This is 170 percent higher than the 1 percent reported hospitalization rate in the 2004 NSS data for the poorest 20 percent of the population. These are still early days in the RSBY program so that utilization rates could end up being even higher as population awareness of its benefits increases. One benchmark is the hospitalization rate observed among the richer groups which is about 4 percent for the richest urban groups in the NSS data. Thus, assumptions on health services use made in Prinja et al. (2012) do seem on the conservative end, at least for poorer groups. And even if their assumptions work in the aggregate, for rich and poor combined, governments may have to allocate greater funds to support health spending, given that public subsidies will focus on the poorer groups.

There are also longer-term trends that point to rising health spending over time in India. Data from the World Health Organization shows that between 1995 and 2009, the annual rate of growth of health spending exceeded the rate of growth of income by 1.3 percentage points in Organization for Economic Cooperation and Development (OECD) countries. Over longer periods, the differences are even more pronounced. Evidence of a rising share of health spending in GDP is also evident for developing countries. Over the period 1995–2009 and with the exception of South Asia (SA), developing countries in all other regions saw rates of growth of health spending per capita that exceeded the rate of growth of income by 0.7 percentage points to 1.5 percentage points annually. In China's case, the difference was 1.1 percentage points annually. Rising levels of underreporting of consumer (and health) expenditure in household surveys likely explains the anomalous finding from India which dominates the South Asian statistics (Bhalla, 2002).

Some of the increase in health spending observed worldwide is due to demographics in the form of rising share of elderly population; and some is due to rising incomes making health-care more affordable. However, existing research points to medical technology, a "catch-all" term used to describe several of the elements that comprise health-care delivery, such as medical devices, drugs, methods of treatment (such as surgical techniques) as the major driver of health spending. Innovations in technology can contribute to rising medical care costs in multiple ways, such as via the introduction of a new and expensive device or drug that replaces preexisting cheaper devices or drugs; by supplementing existing devices or drugs; or by addressing hitherto ineffectively treated conditions. Once introduced, newer and more expensive medical interventions may become more commonly used, thereby adding to higher expenditures and larger budget impact for one set of conditions, even if the intervention is deemed as more cost-effective. Subsequently, such innovations could induce greater attention to hitherto untreated conditions, and a rise in their associated expenditures (Cutler and McClellan, 2001).

Changing medical technology is underpinned by factors on both the demand side (e.g., rising incomes, insurance coverage and greater awareness of newer medical interventions) and the supply side (e.g., provider incentives to offer newer interventions or interventions with higher profit margins).

Rising incomes and better insurance coverage contribute to the adoption of new technology by lowering the economic burden on patients, and use of advanced technology in the private sector may create pressures on government-operated facilities to do the same. This is significant in India, given the dominant role of the private sector in its health service delivery. In the United States, between two-thirds and three-quarters of the increase in health spending has been attributed to technological change and a study for Australia estimated that medical technology changes contributed to about one-third of the growth of health spending during the period from 1992–03 to 2002–03 (Newhouse, 1992; Productivity Commission, 2005). While information on the impact of technological change in India is almost nonexistent, we suspect that the future situation will be no different from that observed in developed countries, given rising incomes and the high potential for off-the-shelf use of technologies already available the developed world. One illustration of this possibility is the data presented in Figure 2 that describes the rapid growth in imports of medical devices in India over the period from 1988 to 2007, following India's liberalization of its import restrictions. Although small in relation to India's GDP, the share of medical device imports (primarily high-tech equipment) quadrupled over this period.

Conclusions

In this paper we examined what it means to provide expanded coverage (including one version referred to as "universal" cover), including financial implications. Achieving a reasonable degree of access to health services to all of India's population is not a trivial task and we have tried to provide some clarity to the implementation challenges that are involved.

In a setting with resource constraints, policymakers will have to decide which population groups they should subsidize and for which services. Efficiency considerations suggest that public subsidies should support services with significant levels of externalities and public good characteristics. Second, equity considerations suggest subsidies for the poor and for children and our potentially disadvantaged groups for health services that result mostly in private benefits. For the nonpoor, efficiency considerations and limited resource constraints suggest government regulatory support for insurance that covers health-care services with mostly private benefits and is funded by the members themselves. Alternatively, if both rich and poor groups belong to the same pool, the richer groups will need to pay

FIGURE 2. India's Imports of Medical Devices, 1998–2007 (at 1999–2000 prices)

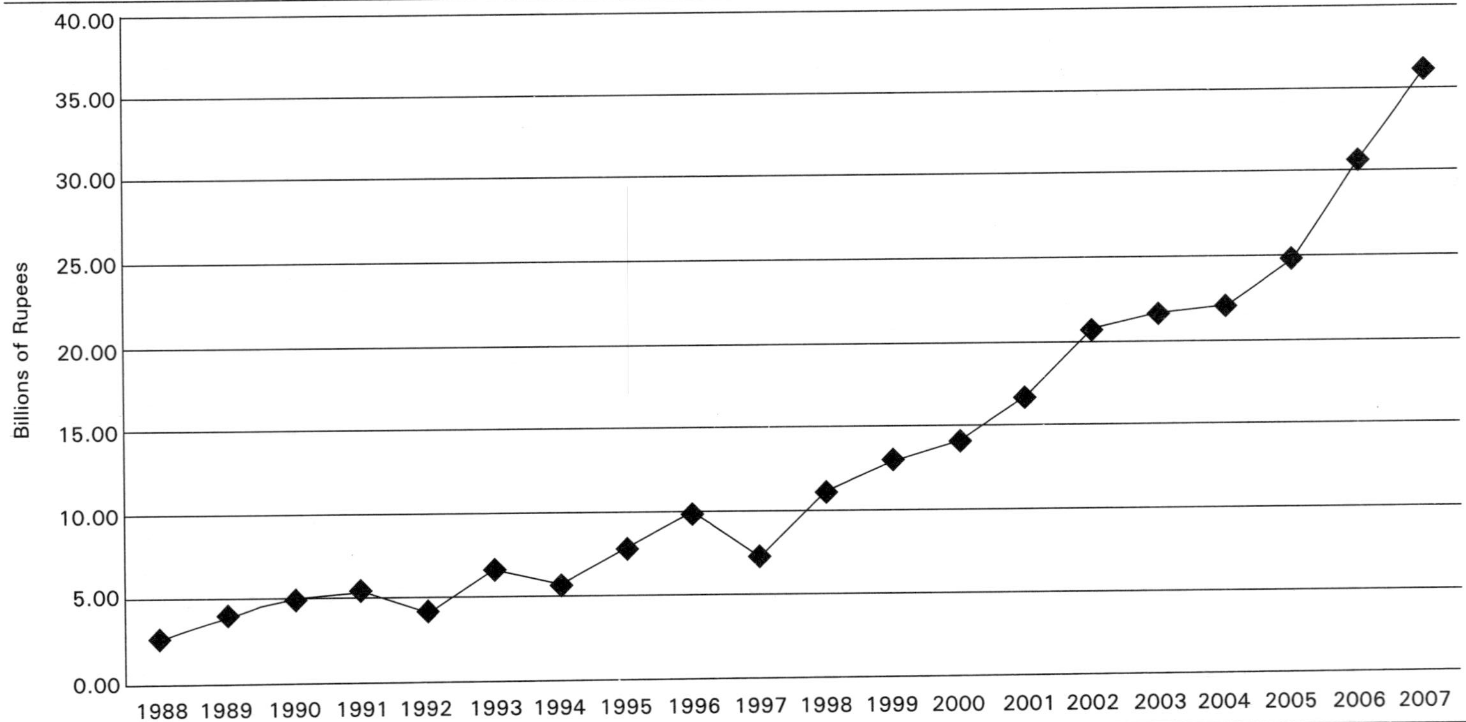

Source: Foreign Trade Statistics of India, Directorate General of Commercial Intelligence and Statistics, Government of India, various years. GDP deflator was used to construct the series at constant prices.

their own premiums. However, there are a number of practical challenges to implementing such a program.

First, in a country such as India with large numbers of preexisting private insurers and health-care providers, considerable regulatory oversight will be required. If richer groups can also buy private insurance coverage, or if their (contribution-based) participation in government-supported insurance programs is voluntary, adverse selection could emerge as a serious problem. Private insurers will offer products that attract good health risks and saddle the public system with sicker patients. This means some thought may need to be given to *compulsory* contributions by nonpoor groups who are allowed to participate in public programs.

Second, given resource constraints, it may not be feasible to instantaneously scale up the program to full coverage in terms of health services covered, as acknowledged in the HLEG report. In this case, policymakers may adopt a strategy of prioritizing those interventions for which the health returns are the highest per dollar, which usually means most prevention and promotion programs. Because financial protection is an important goal as well, Mexico's example serves as a useful guide since it adopted the strategy of prioritizing the most cost-effective interventions for coverage from a separate subgroup of curative services. This also means that considerable research into the costs and cost-effectiveness of interventions will be needed to support policy on an ongoing basis.

Third, policymakers must give considerable attention to the subject of payment mechanisms to address issues of efficiency (and quality). This means increased attention to the capacity of entities that manage financing pools as well as creating an environment where providers respond to payment incentives. Among the key issues that the government will have to address is the degree of autonomy of functioning of public providers, and ensuring fair competition between public and private providers if that is the strategy chosen. The use of provider payment mechanisms, including pay for performance models, have implications for information systems that need to be in place. Possibilities for paying consumers/beneficiaries must also be explored to influence prevention behavior, or to obtain their own outpatient services if issues of fraud or absenteeism in public facilities limit the ability of a funding pool to identify providers in the neighborhood.

Fourth, the state of the supply side—the provision of health services—needs attention. Both the HLEG report and the NCMH report preceding it highlight the need to upgrade health services in the public sector, to create new infrastructure and to contract with private providers. However, considerable challenges remain, particularly in ensuring access to health services in rural areas, and there is a need for better integrating traditional (unqualified) providers who are often the only source of health care for populations in remote areas.

Fifth, India faces serious resource challenges in financing universal coverage, even neglecting long-term trends in demographics, epidemiological shifts, income growth, and technological change that are driving health spending around the world. According to our assessment based on adjustments to estimates presented in the literature, something of the order of 4.5 percent of GDP or higher is required to finance health services for the entire population, of which 1.1 percent will need to come from contributions by households. Thus, levels of government contribution that are considerably greater than anything seen thus far in the health sector are envisaged, including in the projections of the HLEG report. Payroll tax rates on formal sector workers are unlikely to be set at levels that can support health spending of 1.1 percent of GDP, and steps will be required to ensure contributions by nonpoor informal sector workers and other categories of nonpoor, presumably through some form of surcharge on income or wealth.

Finally, we would like to add that health outcomes are only partially influenced by the nature of health financing and delivery, issues that have been the central focus of this paper. The role of clean water, sanitation, and environmental conditions is well known and obviously relevant in the context of morbidity, health-care use, and financial sustainability. A systematic approach to the design of a health-care financing and delivery system ought to model the likely influence of these and other factors as well.

Comments and Discussion

Tapas K. Sen
National Institute of Public Finance and Policy

Ajay's paper is rather a comprehensive one in the sense that it covers a lot of ground and quite a lot of details, and he also actually does not have a very strong preference for any particular combination or any particular solution. What I could make out from the paper is that he does have sympathy for a certain combination, but then he is not very strong on that. That makes it a little more difficult to comment on. But still, what I would do is basically to take up some random issues instead of a structured comment on the entire paper. These are the issues which sort of struck me while reading of the paper; I will simply try to put these issues in the bullet form.

The first thing that I would want to talk about is that there seems to be a presumption in the paper in favor of the insurance system, that we have to have insurance as against the publicly funded, publicly provided kind of system. If we are actually going in for an insurance system mainly and given the fact that already much of the health care is actually provided by private parties rather than public institutions (Ajay is saying at some place that he is expecting the public supply to gradually die out and be replaced by the private providers), then with the combination of an insurance-based system and private providers, we need to worry about the actual price of various kinds of health-care products because experience in several countries, where there is the full insurance system, shows that one does need to worry about it. One has to actually think about really effective methods of controlling the spiraling costs of health care and this is particularly so when one thinks about it in a dynamic sense that examines what happens over the years.

One also has to think about the dynamic stability part if one is talking about a resource-constrained kind of system where the public sector has to at least provide some funding for the entire insurance system. A key issue that has to be settled in a system which is partly funded by the government, partly funded by the individuals, and the combination finances the insurance system and then providers are paid for the health-care products. The key question is: would there be equilibrium in the sense of the costs of the providers, the kind of payments that they would demand, being met—would there be enough funds in the insurance system to actually pay the providers? If you do not have adequate funds then there is a problem because then either you have to cut down on what services you are covering (limited health-care products) or you have to cut down on coverage (subsets of target population). Thus, tricky coverage issues come in particularly when you have a resource-constrained system, and obviously we do have a resource-constrained system in India. We cannot really think of "any amount of cost is okay" kind of situation and, therefore, several issues do get actually linked up: what to cover, who to cover, and also who are the ones that are actually doing the provision of the services. Incidentally, the kind of cost estimates that Ajay comes up with are not very large but at the same time not very small either; the point that I was thinking about is that if we could raise substantive additional resources for the new system, with the same additional resources we could actually beef up the existing system considerably, subject to the constraints which cannot be handled with just money. I suppose anyone would be persuaded to favor the insurance system if it made services available for all, at reasonable costs. But then, one has to think about what is the particular characteristic of the insurance system which really would make it score over the existing system of public supply and also its validity for a country like India. That is something which I think needs to be further discussed, before I am persuaded to accept insurance as the way to go.

There is another issue in terms of coverage: What to cover? Obviously the direct costs are to be covered and Ajay has also talked about some incentive payments for various kinds of services including for children, for healthy habits, and so on. But I would like to add one more consideration in nondirect costs of health expenditure, particularly in the context of India and also particularly in the context of rural people. Because of the lack of access to heath institutions, what happens in rural India is that they actually have to travel far to get proper health care. When that happens, there is a substantive transport cost, there is a substantive cost of people accompanying the patient, they have to stay somewhere when they go out of station, particularly for longer durations; all those costs which are not actually direct costs of health care can be and probably are extremely significant for rural people, and I am not sure whether any insurance system would cover that kind of costs because it would be extremely difficult to make sure how much that cost is and to actually check all those costs. A far better solution obviously would be to provide health care close to everybody, but then that is something which cannot happen just overnight. You need to spend quite a lot of money on the health infrastructure itself—it is not something that is automatically going to take care of itself.

Ajay has mentioned health infrastructure shortages, and has even costed the filling of the gap. At National Institute of Public Finance Policy (NIPFP) we did a series of State studies for various kinds of human development indicators and the kind of costs that are associated with raising those indicators and that kind of thing. Health was one of the human development areas examined. We did some rough estimates of the cost of infrastructure provision to meet the various kinds of norms that are provided in eight individual states. The lessons that we had from those State studies is that, it is first of all very difficult to generalize for the country. There are substantive state-level differences, among other things in the costs that are involved in just providing the infrastructure. I am not talking about other parts of health-care supply, just providing the infrastructure. If I give you the estimated numbers, it does not really help because these were done at different points of time. Thus, one needs to scale the estimates up for population increase, price level, and so on. But let me just say it in a more useful way that for certain states like Maharashtra, West Bengal, and Tamil Nadu, the estimated additional expenditures required to reach the normative levels of supply are not of a scale which could not be met by the respective state governments with a little additional effort and a shift in priorities in favor of health care. But then, there are other states like Chhattisgarh and Orissa, where there is a significant shortage of resources to reach any kind of adequacy in supply of health care. My personal belief is that in the former type of states, because the basic issue of resources is resolved, the systemic issues will also be resolved without much of a problem; in the latter group of states, the basic issue of shortage of resources is not resolved by any of the systems, and hence each of them is as good or bad as another. Then the problem is that we are left with little empirical content in our comparison of one system with another, and thus an informed choice. In fact, I am not quite sure if we need to make a choice for the entire nation. Given the differences between states with respect to several parameters relevant to health care, I could easily plump for a nonuniform health-care system across the country.

One final point on presentation aspect of the paper: Ajay discusses about the systems prevalent in several places, with key details. For uninformed people like me, it would be nice if we could have a brief, tabular comparison of the main features of each of the systems covered, maybe as an Annexure.

Abhijit Banerjee
Massachusetts Institute of Technology

I decided that rather than spend time arguing with the paper, I wanted to see what I knew that might be somewhat useful in answering some of the many important questions it raises. I think that it is a good time to start thinking about this issue. I will say at the end that it is a bad time to close the debate on it or to do very much about it. I think it would be disastrous if we implement Right to Health right now, though that does not mean it will not happen, given the populist pressures that dominate our public life these days.

I think what the paper does very usefully is making clear just how difficult it is to design a universal health system. It is no overstatement to say that it is probably the single largest challenge that any government anywhere in the world faces today. Take the US, which is totally floundering these days, or for that matter, Britain—these are countries with a fair amount of bureaucratic competence that are just sinking under the challenge of designing an effective system of welfare.

Just to illustrate what makes it so difficult, just think of the set of questions that we will need to answer. I am not going to say anything more than what is in the paper, but I will perhaps say it slightly differently and, in any case, it is a useful way to get to what I want to say at the end. First, who should pay for how much of health care? Should we do redistribution through the health system or should people be guaranteed a certain minimum income that would enable them to buy the health care they want at the market price? I think most countries end up redistributing some amount through the health system, but it varies quite a bit across countries. Next, how should we structure the patient's contribution? The patient could either pay some amount every time they use the health-care system or they could pay a lump-sum amount for health insurance and then not have to pay every time they use the system. And should the patient pay a fixed amount irrespective of the amount it actually costs or should he be the residual claimant? Most health systems try to do both: the base payment is fixed per visit, but most health systems allow the patients some form of top-up which allows them to get more than the basic level of services. Patients face a combination of marginal and infra-marginal payments; the incentive implications of the particular structure can be extremely complex.

Competing insurers or single payer? That is another of these issues where there are multiple points of view. I think the recent debate seems to be moving in favor single payers. I do not know that that is necessarily based on much other than the fact that the US (which has competing insurers) is rather disastrous, but the US is unique in many ways. Should the subsidy for health care come from general revenues or from a dedicated pot of money? Should subsidized care be delivered just by government hospitals? Or should private clinics be included as well? When they deliver care, do the hospitals get rewarded at a per customer basis or an FFS? Fees calculated how? You can use DRGs, which is a way to standardize how much treatment for each disease costs, or you can do it on an actual cost plus basis. Should subsidized care be for inpatient treatments only, or outpatient as well? Should there be a cap on total public spending per patient? Or per condition? Or no cap at all? If there is a cap per patient, within that cap do you want to cover all conditions or a fixed list of conditions? These are all independent decisions in a sense. You could have cancer treated or not treated, cancer treated only when it is inpatient, cancer treated both when it is inpatient or outpatient, and each of these decisions has enormous cost implications. And there are many more questions: What kinds of doctors? How you decide who is going to be in charge of the insurance system. And so on.

What do we know that is relevant to the design of such a system for India? Let me say three things that I think we know. One is that we have a huge demand problem. In our data, the average number of visits for very poor people was six visits per adult per year, and the income gradient of visits is actually downward sloping. So, richer people go to the doctor less often than poorer people. These poor patients mostly do not seek treatment for long-term chronic conditions, but palliatives for fever, diarrhea, headaches, etc. That is, things that typically do not need treating, and almost surely a lot of the treatment they are getting for them is either unnecessary or just plain harmful. There is a huge emphasis on injectibles—two out of three visits involves an injection or a drip. Most of the drips tend to be saline or glucose drips which essentially do nothing for anyone except those who are dying from dehydration. There is also massive overuse of antibiotics and steroids which are often taken in partial courses to save money, thus contributing to growing immunities in the bacterial population.

In other words, demand is extraordinarily distorted. This is partly a result of the lack of effectiveness and credibility in our public health-care system. Government doctors are both inept and lackadaisical (Jishu Das' work with many coauthors provides ample proof of this) which then pushes patients into the hands of quacks who promise instant cures and use high-potency drugs to ensure that it happens.

In such an environment, insured outpatient care will be extremely expensive to offer. It will surely be heavily used and might actually exacerbate over-medication. The one potential advantage of going to insured outpatient care—which is probably not big enough to outweigh all of these problems—is that people will be more likely to take the full course of antibiotics if drugs could be made free, rather than just a couple of pills (which helps with the immunity issue). On the other hand, this will probably worsen the problem of unnecessary usage of antibiotics.

There is also a massive supply problem—especially in northern India—as qualified doctors are few and far between. People use quacks—meaning people who do not even claim to have a medical degree. In our data, roughly half the visits are to people who do not claim to have a medical degree. I am sure even fewer of them have a real degree. Yet in recent work in Madhya Pradesh, Das et al. (2011) find that these unqualified doctors outperform the qualified doctors in the government health centers on terms of the efficacy of their treatments (though almost none of them meet the Hippocratic standard of doing less harm than good). However, there is no way the government will be able to empanel these unqualified doctors for publicly funded schemes despite the fact that in many places this would mean excluding the only somewhat competent people who are actually there in the villages from the public scheme. This is not the only issue: There is also the problem of how to deal with the clinics that claim to have qualified doctors who are never physically present and where the treatment is carried out by some hired hack who is just renting the name from the qualified guy. How would the system decide which clinic actually has a qualified doctor and which does not?

Finally, we have clearly created a culture of health care that is profoundly cynical. Most government doctors are absent and practice after hours in government facilities. In this kind of environment where everybody knows that all rules can and have been bent, it is very hard to imagine that there would not be collusion between the doctor and the patient to charge the health system. There is strong evidence of moral hazard on the part of doctors in the US health-care system, leading to something of the order of a 30 percent inflation of costs. If you think of what that means for a context where there is enormous corruption on the ground already, and everyone in the health system is quite cynical, the implications are truly frightening.

Given all that, I think it would be wise to stay away from any system that offers outpatient care. Sticking to inpatient care has several advantages; very importantly, it is easier to regulate. The claim that you are having surgery is relatively easy to check, and also generally means that only specialists are being used; at that level there is much less use of quacks. In fact, we know from the data that for inpatient care people do go to clinics and not to quacks. Indeed, if you look at what poor people do when they get very sick, they go to the government hospital. This is the one part of the government system that is actually used (and maybe some of the community health centers), everything else is essentially empty.

On the other hand, inpatient insurance is a difficult product to sell in many ways. In a field experiment in northern Karnataka, we tried to get people to pay for an insurance product that offered only inpatient coverage; clients found this product very confusing: For example, the insurance only paid for cancer treatment only if there was surgery involved. A client's husband died of cancer, but he never went into the hospital. She spent ₹25,000 on his treatment, but the health insurer refused to reimburse these costs because the insurance was only for inpatient care. The woman, quite naturally, found this very confusing and unfair. In reaction, she and all her friends decided they were going to quit the insurance scheme and, if necessary, quit the MFI that was offering the insurance. More generally, clients really did not like the insurance product. In other words, inpatient care insurance, while much easier to regulate and control, is likely to face an initial lack of demand. This may eventually be overcome by better communication and subsidized pricing (at least until people come to understand the product), but it will take a lot of effort.

Finally, I think the biggest health gains may not be from better health-care funding. The biggest gains are probably in public health and sanitation: getting immunization rates up, getting anemia down, universal access to safe drinking water, and using Oral Rehydration Solution (ORS) for children with diarrhea are probably individually more valuable than anything we can do through the health-care system in the short run. The main case for health insurance is to protect people against the income risk that results from unexpected health-care spending. While this is a very good reason, if I had to pick I would focus on the public health and sanitation issues because they are a lot more straightforward.

General Discussion

T.N. Srinivasan stated that he strongly advocated not going down the rights route. He added that the normative description of what a basket of health care should consist of is not state-specific. You might ask what about the differential abilities of states to provide the normatively defined care. But our Constitution provides an answer to this question. The Finance Commission provides for the transfer of resources across states.

Govinda Rao stated that he was a member of the high-level committee on universalizing health care and that he had been tasked with estimating the cost of universalizing health care. This requires looking at all levels of health care: primary, secondary, and tertiary. The issue is not that the government should provide all health care. Instead, the idea is for the government to provide a package of services in the primary, secondary, and tertiary health care with the package and the norms worked out by doctors and various other people on the Committee. One size does not fit all since there are areas where private sector exists and places where it does not and the State must basically take a decision on whether the package should be publicly provided or you can contract it out. Based on a set of assumptions, Rao placed the cost at somewhere around 2.3–2.4 percent of GDP. This compared with the current public health expenditures of 1.3–1.4 percent of GDP. Therefore, a substantial scaling up of the expenditure needs to be done in the medium term. Rao also expressed some skepticism toward making much of Thai and Mexican experiences. He thought that the type of numbers that will work in the Indian system are very different from those associated with these countries.

Ajay Mahal responded that the Thai population and epidemiological profile is different and to that extent, Thai numbers were not relevant to India. But Mahal said that the paper had scaled those numbers down to account for cost-of-living differences, and so forth, across the two countries. Responding to the questions raised by Srinivasan, Mahal said that he mentioned the rights approach not because he favored it but because that it has been taken by civil society and whether or not we like it, we will have to face up to it. There is a very strong push in favor of the rights-based approach in the health sector in India. Mahal also stated that prevention is important and that ultimately some sort of package will have to be decided upon. It did not have to be a full package but some package of interventions will have to be covered, and over time, if resources permit, one may want to expand it.

References

Akresh, Richard, Damien de Walque, and Harounan Kazianga. 2012. "Alternative Cash Transfer Delivery Mechanisms: Impacts on Routine Preventive Health Clinic Visits in Burkina Faso," Policy Research Working Paper no. 5958, Washington, D.C., The World Bank.

Baeza, Christian and Truman Packard. 2006. *Beyond Survival: Protecting Households from Health Shocks in Latin America*. Washington, D.C.: Stanford University Press and the World Bank.

Banerjee, Abhijit, Esther Duflo, and Rachel Glennerster, 2008. "Putting a Band-aid on a Corpse: Incentives for Nurses in the Indian Public Health Care System," *Journal of the European Economic Association*, 6 (2–3): 487–500.

Bärnighausen, Till and David Bloom. 2009. "'Conditional scholarships' for HIV/AIDS Health Workers: Educating and Retaining the Workforce to Provide Antiretroviral Treatment in Sub-Saharan Africa," *Social Science & Medicine*, 68 (3): 544–51.

Berman, Peter. 1998. "Rethinking Health Care Systems: Private Health Care Provision in India," *World* Development, 26 (8): 1463–79.

Bhalla, Surjit. 2002. *Imagine There's No Country: Poverty, Inequality and Growth in the Era of Globalization*. Washington, D.C.: Institute of International Economics.

Chaudhury, Nazmul, Jeffrey Hammer, Michael Kremer, Karthik Muralidharan, and F. Halsey Rogers. 2006. "Missing in Action: Teacher and Health Worker Absence in Developing Countries," *Journal of Economic Perspectives*, 20 (1): 91–116.

Cutler, David and Mark McClellan. 2001. "Is Technological Change in Medicine Worth It?" *Health Affairs*, 20 (5): 11–29.

Das, Jishnu, Alaka Holla, Michael Kremer, Manoj Mohanan, and Karthick Muralidharan. 2011. "Medical Advice, Quality, and Availability in Rural India." Mimeo, Harvard University.

Das Gupta, Monica, Rajendra Shukla, T.V. Somanathan, and K.K. Datta. 2009. "How might India's public health systems be strengthened?" World Bank Policy Research Working Paper no. 5140, Washington, D.C., The World Bank.

Desai, Sonalde, Amaresh Dubey, Brij Lal Joshi, Mitali Sen, Abusaleh Sharif, and Reeve Vanneman. 2010. *Human Development in India: Challenges for a Society in Transition*. Oxford: Oxford University Press.

Duflo, Esther. 2003. "Grandmothers and Granddaughters: Old age Pensions and Intra-household Allocation in South Africa," *World Bank Economic Review*, 17 (1): 1–25.

Ehrlich, Isaac and Gary Becker. 1972. "Market Insurance, Self Insurance and Self-Protection," *Journal of Political Economy*, 80 (4): 623–48.

Fan, Victoria. 2011. *Essays on Health Systems and Policy in India*, Ph.D. dissertation. Boston, MA: Harvard School of Public Health.

Fan, Victoria, Anup Karan, and Ajay Mahal. 2011. "The Impacts of Aarogyasri Health Insurance on Out of Pocket Health Expenditures in Andhra Pradesh, India," Draft. Washington, D.C.: Center for Global Development.

Fan, Victoria and William Savedoff. 2012. "The Health Financing Transition," Working Paper, Washington, D.C.: Center for Global Development.

Frenk, Julio, Eduardo Gonzalez-Pier, Octavio Gomez-Dantes, Miguel Lezana, and F. Knaul. 2006. "Comprehensive Reform to Improve Health System Performance in Mexico," *The Lancet*, 368 (9546): 1524–34.

Garg, Charu and Anup Karan. 2009. "Reducing Out-of-pocket Expenditures to Reduce Poverty: A Disaggregated Analysis at Rural-urban and State Level in India," *Health Policy & Planning*, 24 (2): 116–28.

Gill, Kaveri. 2009. "A Primary Evaluation of Service Delivery under the National Rural Health Mission (NRHM): Findings from a Study in Andhra Pradesh, Uttar Pradesh, Bihar and Rajasthan," Working Paper no. 1/2009, Planning Commission of India, New Delhi.

Gonzalez-Pier, Eduardo, Cristina Gutirrez-Delgado, Gretchen Stevens, Mariana Barraa-Llorens, Raul Porras-Condey, Natalie Carvalho, Kristen Loncich, Rodrigo Dias, Sandeep Kulkarni, Anna Casey, Yuki Murakami, Majid Ezzati, and Joshua Salomon. 2006. "Priority Setting for Health Interventions in Mexico's System of Social Protection in Health," *The Lancet*, 368 (9547): 1608–18.

Government of India. 2005. *Report of the National Commission on Macroeconomics and Health*. New Delhi: Government of India, Ministry of Health and Family Welfare.

———. 2009. *National Health Accounts 2004–05*. New Delhi: Ministry of Health and Family Welfare, National Health Accounts Cell.

———. 2011a. *Economic Survey*. New Delhi, India: Ministry of Finance.

———. 2011b. *Report of the Expert Committee on National Food Security Bill*. New Delhi: Government of India, Economic Advisory Committee. Available at http://eac.gov.in/reports/rep_NFSB.pdf (last accessed February 29, 2012).

Hawkins, Loraine, Jaruayporn Srisasalux, and Sutayut Osornprasop. 2009. *Devolution of Health Centers and Hospital Autonomy in Thailand: A Rapid Assessment*. Washington, D.C.: HSRI and the World Bank.

High Level Expert Group (HLEG). 2011. *Report on Universal Coverage for India*. New Delhi, India: Planning Commission.

Hsiao, William. 2008. *Health Care Financing: An Overview*. Washington, D.C.: The World Bank Institute.

Hsiao, William, Jonathan Gruber, and Steven Kappel. 2011. *Act 128: Health System Reform Design, Achieving Affordable Universal Health Care in Vermont*. Boston MA: Harvard School of Public Health.

Hughes, David and Songkramchai Leethongdee. 2007. "Universal Coverage in the Land of Smiles: Lessons from Thailand's 30 Baht health reforms," *Health Affairs*, 26 (4): 999–1008.

International Labour Organization (ILO). 2011. "India: State Government Sponsored Community Health Insurance Scheme (Andhra Pradesh)," Bangkok, Thailand: ILO. Available at http://www.ilo.org/public/english/region/asro/bangkok/events/sis/download/paper16.pdf (last accessed on February 29, 2012).

Krishna, Anirudh. 2007. "For Reducing Poverty Faster: Target Reasons before People," *World Development*, 35 (11): 1947–60.

Lagarde, Mylene and Duane Blaauw. 2009. "A Review of the Application and Contribution of Discrete Choice Experiments to Inform Human Resources Policy Interventions," *Human Resources for Health*, Vol. 7, Article 62.

Liu, Xingzhu and Anne Mills. 2003. "The Influence of Bonus Payments to Doctors on Hospital Revenue: Results of a Quasi-experimental Study," *Applied Health Economics and Health Policy*, 2 (2): 91–98.

Liu, Yuanli. 2002. "Reforming China's Urban Health System," *Health Policy*, 60 (2): 133–50.

Mahal, Ajay. 2000. "What Works in Alcohol Policy? Evidence from Rural India," *Economic and Political Weekly*, 35 (45; November 12): 3959–68.

———. 2010. "Health Financing in India," in A. Mahal, B. Debroy, and L. Bhandari (eds), *India Health Report 2010*, pp. 109–26. Delhi and Mumbai: Business Standard Press.

———. Forthcoming. "Malaysia Public Expenditure Review: The Health Sector." Washington, D.C.: The World Bank.

Mahal, Ajay and Victoria Fan. 2010. "The Case for Improving Health in India", in *India Health Report*, pp. 1–20. New Delhi: Business Standard Press.

Martinez, G., N. Aguilera, and D. Chernichovsky. 2009. "The Mexican Healthcare System and the Emerging Paradigm in Modern Systems," Unpublished. Beersheva, Israel: Faculty of Health Sciences, Department of Health Systems Management.

McPake, Barbara and Kate Hanson. 2004. "Implementation of Hospital Reform Policies: Lessons of Experience," draft. London, UK: London School of Hygiene and Tropical Medicine.

Ministry of Health and Family Welfare (MOHFW). 2009. *The National Health Bill 2009*. Available at http://mohfw.nic.in/NRHM/Draft_Health_Bill/General/Draft_National_Bill.pdf (last accessed on June 30, 2011).

———. 2011. *Rural Health Statistics Bulletin: March, 2010*. Available at http://mohfw.nic.in/RHS%20_2010/RHS%20Bulletin-March%202010.pdf (last accessed on June 30, 2011).

Newhouse, Joseph. 1992. "Medical Care Costs: How Much Welfare Loss." *Journal of Economic Perspectives*, 6 (30): 3–21.

———. 1996. "Reimbursing Health Plans and Health Providers: Efficiency in Production Versus Selection," *Journal of Economic Literature*, 34 (3): 1236–63.

Palacios, Roberto. 2010. "A New Approach to Providing Health Insurance to the Poor in India: The Early Experience of Rashtriya Swasthya Bima Yojana," RSBY Working Paper no.1.

Panagariya, Arvind. 2008. *India: The Emerging Giant*. New York: Oxford University Press.

Prinja, Shankar, Pankaj Bahuguna, Andrew Pinto, Atul Sharma, Gursimer Bharaj, Vishal Kumar, Jaya Prasad Tripathy, Manmeet Kaur, and Rajesh Kaur. 2012. "The Cost of Universal Health Care in India: A Model Based Estimate," *PLoS One*, 7 (1): e30362.

Putthasri, Weerasak, Siriwan Pitayarangsarit, Supon Limwattananon, Sripen Tantivess, Roongthong Kharamanond, and Viroj Tangcharoensathien. 2011. "Local Level Responses to Budget Allocations Under the Thai Universal Health Care Coverage Policy," *Journal of Public Health and Development*, 9 (1): 39–54.

Rajasekhar, D., Erlend Berg, Maitreesh Ghatak, R. Manjula, and Sanchari Roy. 2011. "Implementing Health Insurance for the Poor: The Rollout of RSBY in Karnataka." Draft.

Rao, Sujatha. 2012. "Long on Aspiration, Short on detail: Report on Universal Health Coverage," *Economic and Political Weekly*, 47 (6): 12–16.

Reddy, K. Srinath, Vikram Patel, Prabhat Jha, Vinod K. Paul, A.K. Shiva Kumar, Lalit Dandona, for The Lancet India Group for Universal Healthcare. 2011. "Towards Achievement of Universal Health Care in India by 2020: A Call to Action," *The Lancet*, 377 (9767): 760–68.

Rokx, Claudia, George Schieber, Pandu Harimurti, Ajay Tandon, and Aparnaa Somanathan. 2009. *Health Financing in Indonesia: A Reform Roadmap*. Washington, D.C.: The World Bank.

Scheil-Adlung and Florence Bonnet. 2011. "Beyond Legal Coverage: Assessing the Performance of Social Health Protection," *International Social Security Review*, 64 (3): 21–38.

Schneider, Pia. 2010. "Introducing Hospital Autonomy," Seminar on Health Financing and Organization, Moscow: The World Bank.

Shaw, Paul. 2008. "Overview of Provider Payment Mechanisms," Materials for Session 3 of the World Bank Flagship course on October 29, 2008. Washington, D.C.: The World Bank Institute.

Suchonwanich, N. 2010. "Capitation: For Thai Universal Health Coverage." Bangkok, Thailand: National Health Security Office.

Swarup, Anil. 2011. "Rashtriya Swasthya Bima Yojana: Some Initial Trends." New Delhi: Ministry of Labour.

Tang, Shenglan. 2011. "Making universal coverage sustainable in China," Geneva: World Health Organization. Paper presented at the Conference on Universal Coverage in Asia and the Pacific, October 3–4, Bandar Sunway, Malaysia.

Wagstaff, Adam. 2010. "What's the 'universal health coverage' push really about?" November 10. Available at http://blogs.worldbank.org/developmenttalk/what-s-the-universal-health-coverage-push-really-about (last accessed on January 15, 2012).

Wagstaff, Adam, Winnie Yip, et al. 2009. "China's Health System and Its Recent Reform: A Review of Recent Studies." *Health Economics*, 18 (Supplement 2): S27–S23.

Witter, Sophie, Atle Fretheim, Flora L. Kessy, and Anne K. Lindahl. 2012. "Paying for Performance to Improve the Delivery of Health Interventions in Low- and Middle-income Countries (Review)," *Cochrane Systematic Review* 2.

World Bank. 2011. *Restructuring Paper on a Proposed Project Restructuring of Health Insurance Project*, Report no. 62850. Washington, D.C.: The World Bank.

World Health Assembly. 2005. "WHA 58.33: Sustainable Health Financing, Universal Coverage and Social Health Insurance." Geneva: World Health Organization. Available at http://apps.who.int/gb/ebwha/pdf_files/WHA58/WHA58_33-en.pdf

World Health Organization (WHO). 2008. "World Health Statistics 2008." Available at http://www.who.int/whosis/whostat/2008/en/index.html (accessed on June 20, 2010).

———. 2010. *The World Health Report: Health Systems Financing—the Path to Universal Coverage*. Geneva: WHO.

Xu, Ke, Priyanka Saksena, and Alberto Holly. 2012. "The Determinants of Health Expenditure," Working Paper, Results for Development Institute, Washington, D.C.

Yip, Winnie and Ajay Mahal. 2008. "Health Care Systems of China and India: Performance and Future Challenges," *Health Affairs*, 27 (4): 921–32.

Yip, Winnie and Karen Eggleston. 2001. "Provider Payment in China: The Case of Hospital Reimbursement in Hainan Province," *Health Economics*, 10 (4): 325–39.

———. 2004. "Addressing Government and Market Failures with Payment Incentives: Hospital Reimbursement Reform in Hainan, China," *Social Science and Medicine*, 58 (2): 267–77.

Yip, Winnie and William Hsiao. 2009. "Non-evidence Based Policy: How Effective Is China's New Cooperative Medical Scheme in Reducing Medical Impoverishment?" *Social Science and Medicine*, 68 (2): 201–09.

SHIKHA JHA
Asian Development Bank

BHARAT RAMASWAMI
Indian Statistical Institute

The Percolation of Public Expenditure: Food Subsidies and the Poor in India and the Philippines*

ABSTRACT The paper measures the percolation of food subsidy expenditures to the poor by defining a metric that takes into account the depth and width of income transfer. The metric is applied to food subsidy expenditures in India and the Philippines, which operate similar in-kind transfer schemes. The emphasis in the literature has been on reducing inclusion errors. While the metric takes this into account it also captures nontargeting leakages coming from excess costs and fraud. The principal finding is that percolation is poor in both countries and the payoffs from reducing nontargeting leakages are large relative to lowering inclusion errors.

Keywords: *Food Subsidy, In-kind Transfer, Targeting, Excess Cost, India, Philippines*

JEL Classification: *H22, I38, Q18*

* This paper is written for the 2011 India Policy Forum. We are deeply grateful to Surjit Bhalla, David Coady, Bhaskar Dutta, Reetika Khera, Aashish Mehta, Rinku Murgai, Shekhar Shah, P.V. Srinivasan, and participants of the India Policy Forum for valuable comments. We thank Siraj Hussain, formerly of the Ministry of Consumer Affairs, Food and Public Distribution, Government of India and currently Managing Director, Food Corporation of India, for facilitating access to data about state-level offtake of subsidized foodgrains. We gratefully acknowledge the superb research assistance of Pilipinas F. Quising and Ronald Tamangan. The usual disclaimer applies.

Introduction

Imagine the following thought experiment. Suppose there is to be a marginal expansion of a food subsidy program. What would be the impact of this policy on the poor? This paper provides some answers to this question for the food subsidy programs of India and the Philippines. The paper puts forward a metric that lies between zero and one. If the metric is close to one, it indicates that most of the expenditure from a marginal expansion of the subsidy program percolates to the poor. On the other hand, if the metric is close to zero, it indicates the poor receive very little benefit from a marginal expansion.

Indian economic growth in the 2000s has been in the high single digits and has catapulted the economy into the ranks of the best global performers. The Philippines economic growth has been steady for most of the period. Although deprivation is far greater in India, poverty is a serious problem in both countries and there is debate on whether economic growth has sufficiently trickled down to the poor. In India, there is pressure on policymakers to invest the tax dividends of economic growth on safety net and social sector programs. In both countries, the efficacy of existing antipoverty programs is a continuing concern. These reasons justify the question posed in the preceding paragraph.

Both India and the Philippines expend significant resources on food subsidies through in-kind transfer programs. As will be discussed later, there are many similarities between the two programs. There are some notable differences as well—the most important of which is that the Philippine program is not targeted unlike the Indian program. The value of a comparative analysis between India and the Philippines lies arguably in identifying generic issues with in-kind transfer schemes that determine the extent to which food subsidy expenditures percolate to the poor.

We follow the literature in quantifying the benefits to households in terms of income equivalents, i.e., the implicit income subsidy that is equal to the product of the quantity purchased of the subsidized commodity and the difference between the market and subsidized price (Besley and Kanbur, 1993; Coady et al., 2004). The academic and policy literature recognizes that the gains to the poor depend on targeting as well as program delivery. So if we think of the flow of resources from government coffers to poor households, then we have to think of how it percolates through the claims of nonpoor households and of stakeholders that implement the subsidy program.

However, most of the studies have only evaluated the targeting performance of subsidies. From this literature, it is well known that most transfer programs are costly because of substantial nontarget beneficiaries. For instance, from a survey of universal food subsidy schemes, Coady (2002) finds that the median targeting performance implied that the government spent $3.40 to transfer $1.00 to the poor. In their meta-survey of income transfer programs, Coady et al. (2004) conclude that interventions that use some methods of targeting (e.g., means testing, geographic targeting, or self-selection in public works) result in the target group receiving a greater share of benefits. Further, a standard policy prescription, especially from multilateral institutions, is to recommend that governments should target subsidies toward the poor and not waste resources subsidizing the nonpoor.

However, there is no generalized theoretical presumption that policy should always aim to reduce inclusion errors. The literature offers examples where targeting is costly both administratively as well as in economic terms because of incentive effects (Besley and Kanbur, 1993; Kanbur, 2009). In addition, Gelbach and Pritchett (2000) argued that programs that are tightly targeted toward the poor (i.e., low inclusion errors) do not receive political support from the nonpoor, and thus are ultimately endangered. In addition, there are the practical difficulties of targeting.

In their meta-survey of studies that evaluate income transfer programs, Coady et al. (2004) found very few studies that looked at how program costs affect the percolation of benefits. And even such information consisted only of administrative costs ignoring the costs due to corruption or theft. In this paper, we quantify the extent to which food subsidy expenditures percolate to the poor taking into account targeting leakages as well as leakages due to deficiencies in program delivery that result in excess costs and fraud.

Our principal finding is that the payoffs to program delivery that reduces waste are much larger than the gains from lower inclusion errors. While opportunities for reducing such targeting errors exist in both India and the Philippines, the payoffs from such policies are distinctly secondary to the payoffs from reduction of waste. We shall argue that such a finding is important because reducing inclusion errors is not only contentious politically but is also a policy recommendation that is accompanied by many caveats in the economics literature. On the other hand, it is straightforward to recommend policies that deliver subsidies more efficiently. Indeed, as we shall see, higher percolation may well require greater tolerance of inclusion errors.

Program Description

India and the Philippines operate food subsidy programs, referred to in this paper respectively by their acronyms TPDS or Targeted Public Distribution System (which is spearheaded by the Food Corporation of India [FCI], a body under the central government) and the NFA or National Food Authority (which too is a central government owned and controlled corporation tasked to ensure food security of the country). The commodities that are subsidized in these programs include staple foodgrains. The Philippines program subsidizes mainly rice while the Indian program offers subsidies on rice and wheat.[1] One key difference is that the TPDS primarily sources grain from domestic procurement while the NFA program depends heavily on imports (over which it has a monopoly). Table 1 presents a descriptive summary of the programs in these two countries, which have similar mandates and many commonalities in functioning as well. The mandates are multiple, including price stabilization, ensuring food access by the poor and supporting farm prices. The two programs are thus expected to balance producer and consumer interests in each country. The commonality in functioning is that both these programs deliver in-kind subsidies. Because of in-kind subsidies, both countries have government agencies that source, store, transport and distribute the grain to designated retail outlets.

In the Philippines, apart from its monopoly of rice imports, the NFA procures *palay* or paddy rice from farmers and farmer organizations at a relatively high price compared to the market farm price. The NFA also carries out other activities such as buffer stocking, processing, dispersal of *palay* and milled rice to strategic locations, and distribution to various marketing outlets through the year. To assist consumers, the NFA sells rice through accredited retailers at a mandated, below-market price. The retailers receive a fixed margin on the sale.

In India, the central and state governments together run a marketing channel solely devoted to the distribution of the subsidized food. At the retail level, this involves a network of "Fair Price Shops" (FPS) which sell subsidized grain to consumers. Subsidized grain is not accessible elsewhere. The FPS is usually run by private agents who receive a fixed percentage as commission for their sales. The FPS is often restricted to sell only subsidized grain. The central government is responsible for procurement, storage, transportation and bulk allocation of foodgrains to different states. The state government is responsible for transporting and distributing the grain within the state through the network of FPS.

The NFA rice subsidies are universal with supposedly unlimited purchase. However, there are exceptions—within the NFA program is a smaller program called Tindahan Natin Program (TNP). This program operates

1. While these programs also subsidize other consumption goods, we focus on these staples as they account for a major share of the subsidies.

TABLE 1. A Comparative Summary of Food Subsidy Programs in India and the Philippines

Program design and functioning	*India*	*Philippines*
Main staple commodities	Rice and Wheat	Rice
Volume of grain distributed	32 million tons (2004–08)	1.6 million tons (2004–07)
Targeting	Yes—at household level.	No. Universal program with small targeted programs
Quota	Yes. Fixed per household.	No. Unlimited quantities.
Subsidized price	Yes.	Yes.
Source of supply	Domestic procurement—supplemented by imports in exceptional years.	Largely imports (rice) supplemented by domestic procurement.
Operations	Supply from central government to state warehouses by FCI. Supply from state warehouses to ration shops by state governments.	Supply from central government to NFA warehouses to accredited and licensed private retail outlets and institutions and government rolling stores.
Funding	Central government budget.	Central government budget. Official Development Assistance to the Philippine government. Loans from the public and private sectors.
Budgetary allocations as percentage of GDP	0.72 percent (2004–07)	0.3 percent (2005–08)

Sources: Economic Survey, Government of India; National Food Authority Accomplishment Reports (NFA, various years); National Food Annual Audit Reports (Commission on Audit, various years); CEIC Data Company Ltd., accessed January 8, 2010; author's computations.

through dedicated outlets that sell only the NFA-subsidized commodities. The program is supposed to favor the setting up of these stores in the poorer regions through geographical targeting. Since 2008, individual-based targeting is also being attempted. In this experiment, which is confined to Metro Manila, the target beneficiaries are families with incomes less than PhP 5,000 per month. Such identified households are eligible to 2 kg of rice at subsidized prices.

Despite its universal nature, household expenditure survey (Family Income and Expenditure Survey or FIES) data for 2006 indicates that out of 12 million households, only about 2 million purchase rice, i.e., about 16 percent of the population. One reason for this could be self-targeting through inferior quality. According to World Bank (2001), the NFA mixes good quality rice with poor quality rice for most of its releases. Moreover, retailers may mix the NFA releases of any good quality rice with bad quality rice. Another reason could be the unavailability of the NFA rice in some parts of the country.

India moved from a general entitlement scheme (which was widely criticized for its failure to serve the population below the poverty line) to targeted food subsidies in 1997. The current regime is therefore called TPDS. Subsidies depend on whether the household is classified as APL, BPL, or POP or the *Antayodaya Yojana* program.

All households are entitled to a monthly quota of 35 kg of rice or wheat per month.[2] In principle, the prices of subsidized grain are supposed to be fixed with reference to the government's "economic cost," i.e., the cost incurred by government agencies in procuring, storing, transporting, and distributing grain. BPL households are supposed to receive 50 percent subsidy (i.e., 50 percent of economic cost) while APL households are not supposed to be eligible for any subsidy at all.[3] The prices for POP households are fixed below that of BPL households and not with reference to economic cost.

Table 2 lists the price of rice and wheat for each category of households and also the economic cost for the most recent years. The subsidized prices in Table 2 were fixed in 2002 on the basis of the principles outlined in the previous paragraph. However, these prices have not yet been subsequently revised. As a result even the APL households in 2008/09 received a subsidy in excess of 50 percent of economic cost. The qualification to this is that the

2. Some states (e.g., Andhra Pradesh, Rajasthan, Karnataka, Tamil Nadu) combine limits below 35 kg with lower prices or expanded coverage or both (Khera, 2011).

3. In practice, as we shall see later, even APL households receive subsidies and the subsidy to BPL households has exceeded the 50 percent benchmark.

TABLE 2. Subsidized Price of Rice and Wheat in India According to Household Type, 2009 (₹/kg)

	POP	*BPL*	*APL*	*Economic cost (2007–08)*	*Economic cost (2008–09)*
Rice (Common Variety)	3	5.65	7.95	15.64	17.9
Wheat	2	4.14	6.10	13.53	13.93

Source: Economic Survey, Government of India.
Note: BPL = below poverty line, APL = above poverty line, POP = poorest of the poor.

central government does not guarantee full supply to the state governments for its APL requirements. The actual allocation depends on past purchases and ad hoc considerations. The total number of households within a state that are eligible to be classified as BPL is made through an expenditure sample survey administered by the central government.[4]

The list of BPL beneficiaries is prepared through a BPL census. In the latest census of 2002, households received scores based on 13 criteria. The BPL households were identified as those who fell below a cutoff score (which was decided by the respective state governments). If the total number of BPL identified households exceeds that which is estimated by the central government, the subsidy on the excess households has to be borne by the state government.

Both India and the Philippines expend significant resources in operating their food subsidy programs. In the case of India, the budgetary cost of food subsidy topped 1 percent of GDP in 2002 but later came down to around 0.72 percent toward the later part of the decade. The decline happened because of the rapid growth in GDP since about 2003. The Philippines program is heavily dependent on imports, and so the cost of the program varies with world prices. The program cost averaged 0.3 percent of GDP between 2005 and 2008 (Table 1). Because of high world prices for food in 2008, the program absorbed 0.6 percent of GDP that year.[5]

4. The initial estimates of the state-wise BPL population was done for 1993/94 as the product of (*a*) the estimate of the proportion of households that are poor in 1993/94 and (*b*) the total population in 1995. The latter has since been revised to 2000; however, the former estimate has not been revised yet.

5. Since this study was conducted, it has been decided that the NFA's subsidized distribution program will be gradually phased out (Javier, 2011). NFA's budget has been trimmed down in the light of several inefficiencies that have been highlighted in recent years. Bulk of its operational funds has been reallocated to the Department of Social Welfare and Development to finance its conditional cash transfer program.

Impact of Food Subsidies on the Poor: A Measure of Percolation

If public expenditure on food subsidies increases marginally, how much of it percolates to the poor? This section posits a measure of percolation. The starting assumption is that there is a clear classification of households into the poor and nonpoor.

The simplest way to examine a program for its effectiveness in reaching the poor is to consider its exclusion and inclusion errors. Let r denote the rate of participation of the poor, i.e., the proportion of the poor who participate and receive benefits from the subsidy program. $(1-r)$ is the proportion of the poor who do not receive food subsidies. It is called the *exclusion error*. The *inclusion error* is defined as the proportion of subsidy recipients who are not poor. A subsidy regime is said to be targeted well if both these errors are low.

Targeting effectiveness is, however, a limited measure of percolation. In particular, inclusion errors only tell us about how many recipients are nonpoor but not how much subsidies they get. This problem can be rectified by considering the share of the poor in the subsidy. This is denoted by s. It captures the *depth* of percolation (i.e., the extent to which expenditures reach the poor).

s is the targeting measure that is used most widely in studies evaluating income transfer programs and was therefore used by Coady et al. (2004) to compare targeting effectiveness across programs in a meta-survey of different studies. This measure can also be justified as the social valuation of income transferred to poor households, when poor households receive a welfare weight of unity and nonpoor households receive a zero welfare weight (Coady et al. [2004]). s is negatively related to the inclusion error (Ravallion [2009]). Quite clearly, if the inclusion error is zero then the poor receive the entire subsidy.[6] At the other extreme, if the inclusion error is 100 percent, then the fraction of the subsidy reaching the poor is zero. It can also be shown that s captures the impact of a program on the poverty gap per unit of public spending provided that the program does not by itself change the head count measure of poverty and if there are no fiscal costs other than transfers (Besley and Kanbur, 1993; Ravallion, 2009).

For a marginal expansion of public expenditures on food subsidy, we assume that the share of the poor in the incremental subsidy is the same

6. The statement assumes that the entire subsidy is spent on income transfers. If, for instance, some of the subsidy is spent on administrative costs, then the share of subsidy going to the poor is less than one even when there are no inclusion errors.

as the average share, i.e., the marginal share is equal to the average share. In this case, s can also be a measure of percolation of public expenditures. It measures the income transfer to the poor for a unit expansion of public expenditures.

The share measure is, however, insensitive to the *width* of percolation (i.e., the coverage among the poor). We could have a well-targeted program with high s, but the program may yet have modest impacts on incomes of the poor because of exclusion errors. For instance, suppose $s = 1$ and imagine two scenarios. In scenario A, only 10 percent of the poor receive subsidies. In scenario B, subsidies are accessed by 50 percent of the poor. A unit expansion of public expenditures will lead to a wider percolation in scenario B (together with lower per capita individual gains for those receiving subsidies) than in scenario A. The share measure cannot accommodate a preference for scenario B over scenario A.

To motivate our metric of percolation, suppose the policymaker has a welfare function with the following properties: (*a*) the function is invariant to the incomes of nonpoor households; (*b*) the function is increasing in the incomes of each of the poor households; and (*c*) for a given transfer to the poor, the welfare function is increasing in the coverage among the poor, i.e., greater is the number of recipients, higher is the value of the function.

The share measure is consistent with the first two restrictions but not with the third. A percolation measure consistent with all the three restrictions is $Y = rs$ where r is the percentage of the poor that participate in the food subsidy program. The measure Y lies between zero and one. If either of s or r is zero, then the metric is zero as well. Similarly, the maximum value of Y is one which happens when all of the poor participate and when they receive all of the subsidies. When the participation rate is 1, the percolation metric reduces to s.

The value of our percolation measure is that it combines the depth and width of income transfer into a single index. However, the metric is not perfect. All nonpoor households have zero weight in the welfare function, irrespective of their distance from the poverty line. The metric also does not explicitly distinguish between poor households, except to the extent that s incorporates the program impact on the poverty gap for every unit of program budget. It treats s and r symmetrically. For instance, a transfer to the poorer half of the poor population would be valued the same as the transfer that is half of this magnitude but to all of the poor.[7] As we shall see, however, a

7. We owe this observation to David Coady.

finer measure of percolation, while desirable, would not materially alter the findings of this paper.

Computing *s*—The Fraction of Subsidy Received by the Poor

Inclusion errors mean that if a government spends $1 on provision of food subsidy, poor households receive only a fraction of it. Such a diminution in the amount of subsidy that reaches poor households is called a *targeting leakage*. While it is generally agreed that a targeting leakage (due to inclusion errors) should be minimized, the debate in the income transfers literature is whether and how it can be done. The debate is enduring because minimizing inclusion errors can be costly (administratively) and often leads to greater exclusion errors. With such a trade-off, optimal targeting depends on how much weight the government puts on inclusion error relative to exclusion error.

However, there can also be other sources of leakage. In particular, the subsidy received by all households is often less than the expenditure incurred by the government. In this section, we argue that s—the fraction of subsidy received by the poor also ought to be adjusted for nontargeting leakages. There is agreement in the literature on this requirement (Besley and Kanbur, 1993; Coady, 2002), but is generally ignored usually because of lack of data.

Let p be the market price of the food staple and let k be its subsidy price. If q is the total quantity consumed of the subsidized staple, then the income subsidy received by consumers is

$$I = (p - k)q \tag{1}$$

The government's cost of food subsidy is denoted by C and it can be written as

$$C = (a - k)Q \tag{2}$$

where a is the government's cost of acquisition and distribution of the food staple and Q is the total supply of subsidized staple that is distributed by the government. Then C can be decomposed as

$$C = ((a - p) + (p - k)Q) = (a - p)Q + (p - k)\,(q + d) \tag{3}$$

where $d = (Q - q)$ measures the government supplies for distribution that never reach households through the subsidy mechanism. These represent the illegal diversions by intermediaries that profit from arbitraging the difference between the market and subsidy prices. Hence, we have

$$C = (a - p)Q + (p - k)q + (p - k)d = I + (a - p)Q + (p + k)d \qquad (4)$$

In this analysis, the income subsidy received by all households I is less than the government's cost of providing subsidies because of two components. The second component $(a - p)Q$ on the right hand side of equation (4) reflects the difference between the government's cost of purchase and distribution of grain and the price in the market. We call this *excess* cost. This can arise either because the government buys the food staples at higher prices than the private sector (e.g., as a result of price support operations) or because the government is inefficient relative to the private sector or because of a combination of these reasons. The third component $(p - k)d$ is the cost of illegal diversions.

Finally, I itself can be broken up into two components: the income transfer to the poor (denoted as I_p) and the income transfer to the nonpoor group (denoted as I_n). Hence we can write (4) as

$$C = I_p + I_n + (a - p)Q + (p - k)d \qquad (5)$$

The fraction of government expenditure received by the poor is therefore

$$s = I - [(I_n/C) + ((a - p)Q/C) + ((p - k)d)/C] \qquad (6)$$

s is the difference between one and the sum of three kinds of leakages. The first leakage is the targeting leakage, the second source is the leakage due to excess costs and the third leakage is because of illegal diversions of the subsidized staple to open markets. In the sections that follow, we report on estimates for each of these leakages for India and the Philippines and the cumulative outcome for s.

Targeting Errors

In what follows below, we define the poor as that part of the population that subsists on expenditures below the official poverty line.

Philippines

The distribution of NFA rice is not targeted. Hence it should be possible in principle to achieve zero exclusion error. Yet, only 25 percent of the poor received benefits from the subsidy in 2006 (see Table 3). This is a modest improvement over the situation in 2003 where only 20 percent of the poor participated in the program. Thus the exclusion error of the program continues to be large.

Table 3 also considers the poor/nonpoor composition of the population that receives NFA rice. Of the beneficiaries in 2006, 52 percent are poor while 48 percent are nonpoor. Thus, it would seem that the inclusion error is also large even though there has been some improvement from 2003.

TABLE 3. Exclusion and Inclusion Errors of the NFA Program

Year	*Participation rate*	*Exclusion error (percentage of poor who are excluded)*	*Inclusion error (percentage of recipients who are nonpoor)*
2006	24.5	75.5	48.3
2003	20.2	79.8	56

Source: Computed from Philippine Family Income and Expenditure Surveys.

Comparing urban and rural areas, the exclusion error is equally large (about 75 percent) in both urban and rural areas (Figure 1). In 2006, the participation rate of the poor was 24.6 percent in the rural sector and 24.2 percent in the urban sector. However, the inclusion error is more serious in urban areas than in rural areas. Figure 1 shows that in urban areas, as many as 68 percent of beneficiaries are nonpoor as against 39 percent in rural sector. The ease of access to NFA-accredited retailers, the better supply of NFA rice and lower opportunity costs for the urban rich (who can send household domestics to queue up for NFA rice) may be factors that contribute to higher purchases of NFA rice by the urban nonpoor.

Inclusion errors may not be consequential if the nonpoor recipient households buy very little NFA rice. To assess this possibility, consider Figure 2 which presents the per capita consumption of NFA rice among poor and nonpoor recipients. It shows that both poor and nonpoor recipient households buy about the same quantities of NFA rice on per capita basis. This suggests that inclusion errors are indeed serious. As annual per capita grain consumption varies from 90 (for the poorest decile) to 140 kg (for the richest households), NFA rice accounts for more than 50 percent of the rice consumption of poor recipient households and more than one-third of the rice consumption of nonpoor recipient households.

FIGURE 1. Targeting Errors, by Sector of Residence

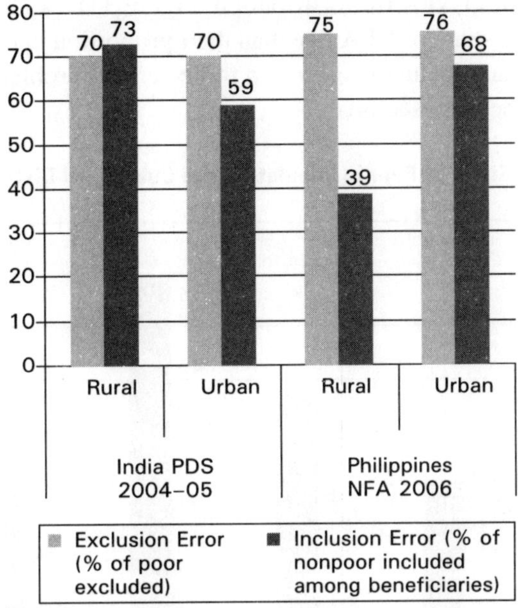

Source: Computations from Expenditure Surveys.

FIGURE 2. Quantity of Subsidized Grains Purchased (kg per capita per year)

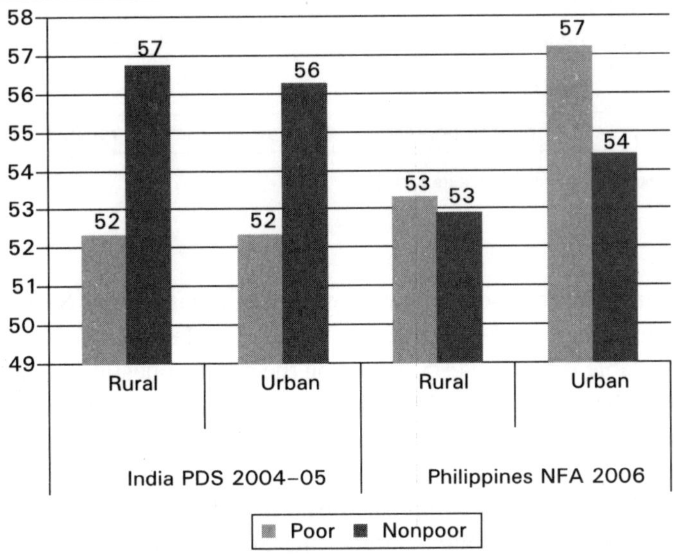

Source: Computations from Expenditure Surveys.

A more comprehensive measure of inclusion errors is to consider the share of the poor in NFA rice distribution. Figure 3 shows that the poor do receive a greater share of NFA rice than their proportion in population. The table confirms that inclusion error is a more serious problem in the urban sector than in the rural sector.

FIGURE 3. Share of Poor in Population and Subsidized Distribution

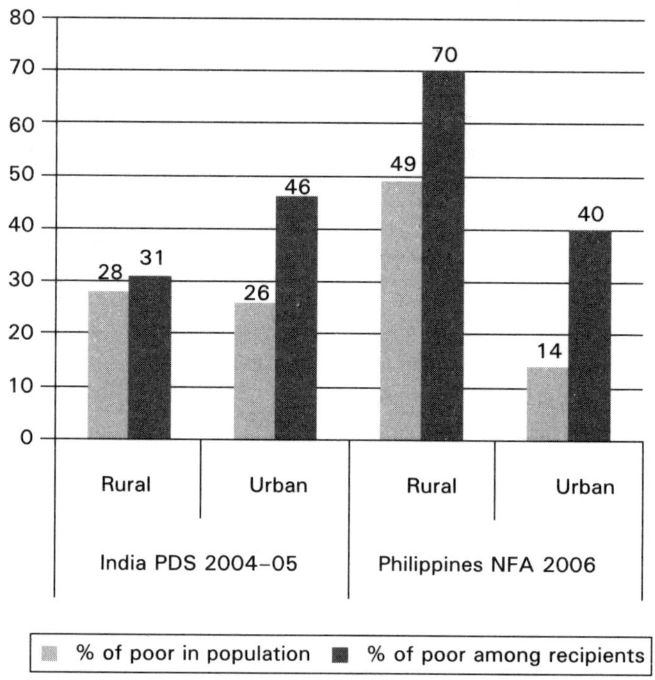

Source: Computations from Expenditure Surveys.

India

The consumption expenditure survey of the National Sample Survey (NSS) provides information about targeting errors. The latest large-scale survey that is available is for 2004/05. Based on the survey questions, a household is defined to be a recipient of food subsidies if it purchases subsidized rice or wheat or both during the survey reference period. While the targeted public distribution system (PDS) was launched in 1997, it is generally agreed that targeting was not fully accomplished by 1999. Therefore, the results from 1999/2000 (when the previous large-scale expenditure survey was carried out) may be seen as corresponding to a pre-targeting regime while those from 2004/05 refer to a targeted subsidy regime.

Table 4 compares targeting errors from 1999/2000 to 2004/05. The table shows a rise in exclusion error and a fall in the inclusion error. However, the changes are small. In 1999/2000, the program was not well targeted. This situation does not change in 2004/05 despite the introduction of targeting in the design of the program.

TABLE 4. Exclusion and Inclusion Errors of TPDS

	Participation rate	*Exclusion error (percentage of poor who are excluded)*	*Inclusion error (percentage of recipients who are nonpoor)*
2004/05	30	70	70
1999/2000	36	64	76

Source: Computed from the Expenditure surveys of the National Sample Survey.

Figure 1 compares exclusion and inclusion errors across urban and rural areas. Exclusion errors are uniformly high at 70 percent in both sectors while the inclusion errors are higher in rural areas.

Exclusion errors could happen either because households chose not to participate in the program or because of mis-targeting.[8] As mentioned earlier, targeting is based on proxy indicators that are elicited from a household census. Mis-targeting could happen in two ways. First, a poor household may not be classified at all. In this case, the household does not receive the food eligibility card[9] and cannot make purchases from the PDS. Second, even if a household receives a food eligibility card, it may be wrongly classified as an APL household and is not therefore entitled to the larger subsidy offered to households classified as BPL or POP. The consumption expenditure survey reports whether households possess food eligibility cards and of what type.

Let N be the number of poor households. We divide this into three categories: N_1, the number of poor households that do not possess a food eligibility card; N_2, the number of poor households that are classified as APL; and N_3 the number of poor households that are classified as either BPL or POP. Let b_i, $i = 1, 2, 3$ be the number of poor households that purchase food from the PDS in each of these three categories, respectively. If b is the total number of poor households that purchase food from the PDS, the participation rate of the poor can be written as

8. Households might not participate because of various reasons such as low quality of publicly provided grain, distance to retail outlets, unavailability of supplies or lack of liquidity.

9. The food eligibility card is popularly referred to as a "ration card" in India.

$$r = (b/N) = (b_1/N_1)(N_1/N) + (b_2/N_2)(N_2/N) + (b_3/N_3)(N_3/N) \quad (7)$$

Equation (7) expresses the overall participation rate as the weighted sum of participation rates of the poor in each of the three categories, with the weights being the proportion of the poor in each of the three categories. Notice that the proportion of the poor in categories one and two is an evidence of mis-targeting.

Table 5 displays the conditional participation rates and the associated weights for the rural and urban sector. Consider first the rural sector. For poor households that hold either the BPL or POP eligibility card, the participation rate is 61 percent. This drops sharply to 13 percent for households with APL eligibility.[10] For households without any eligibility, the participation rate is 4 percent.[11] The associated weights are 0.4, 0.4, and 0.2, respectively. In other words, 60 percent of the poor are either classified incorrectly as APL or not classified at all (i.e., without eligibility to any subsidy).

TABLE 5. Decomposition of Participation Rate of the Poor in TPDS

	Rural			*Urban*		
Category	*Conditional participation rate I*	*Proportion of poor II*	*Unconditional participation rate III = I × II*	*Conditional participation rate I*	*Proportion of poor II*	*Unconditional participation rate III = I × II*
No card	0.04	19.57	0.86	0.03	27.83	0.92
APL	0.13	40.52	5.27	0.18	44.83	8.05
BPL+POP	0.61	39.90	24.51	0.77	27.34	20.94
Sum	–	100.00	30.64	–	100.00	29.91

Source: Computations from the Expenditure surveys of the National Sample Survey.
Note: APL = above poverty line, BPL = below poverty line, POP = poorest of the poor.

If this kind of mis-targeting is eliminated and all poor are classified as either BPL or POP, the participation rate would improve. If the participation conditional on eligibility remains invariant, then the participation rate would

10. Given that price for APL cardholders is less than the market price, it may seem somewhat of a puzzle that poor households with such eligibility have such low participation rates. The reason for this outcome is that the central government does not guarantee full grain allocations to the states against the APL category. The actual allocation depends on past purchases and ad hoc considerations. The unpredictable supply to APL grain cardholders dampens their participation.

11. Households without eligibility might still access subsidized food supplies using the ration card of others.

nearly double from 31 percent (total unconditional participation rate) to 61 percent (conditional participation rate for BPL and POP categories) in the rural sector. Hence mis-targeting is a major reason for the high exclusion error. Notice, however, that participation does not reach 100 percent because nearly 40 percent of poor households do not participate despite eligibility. This underscores that there are factors other than eligibility that are also barriers to participation. The analysis for the urban sector is similar: here the gains from correct targeting are greater as the participation rate would rise from 30 percent to 77 percent.

If households received subsidized grain, how much did they receive? This question is answered in Figure 2 which displays across poor and nonpoor households the amount of grain purchased through TPDS. Figure 2 shows that the extent of use does not vary between poor and nonpoor households. Per capita grain consumption from all sources varies between 10 and 12.5 kg per month for poor and nonpoor households. Thus, the TPDS on average accounts for about 40 percent of total grain consumption of the households that receive subsidies. Note also that for an average family of five, total household monthly consumption is nearly 20 kg, which is much less than the entitlement of 35 kg per month.

Figure 3 presents the share of poor in total grain quantity distributed through the TPDS.[12] This is compared to the share of the poor in total population. Although the quantity share is greater than the population share, the poor receive less than 50 percent of the total quantity distributed.

Leakages (Due to Illegal Diversions)

Because of the price difference between subsidized grain and grain sold through regular marketing channels, there are powerful incentives to arbitrage and make illegal profits. Both countries have various audit and inspection systems to police such theft. Leakages are the illegal diversions of subsidized grain to regular market channels.[13] They are typically estimated by comparing the distribution of subsidized grain from administrative records to the receipt of grain by households calculated from survey data.

12. The total quantity distributed through TPDS is computed from the household expenditure survey. It is not the total quantity of grain supplied to the TPDS by the government.

13. Sometimes leakages are also used to refer to the receipt of subsidized grain by non-target groups. This is a leakage due to targeting error. In this section, we are concerned with leakages due to corruption and fraud.

For the Philippines, Mehta and Jha (2009) report a 54 percent gap between the NFA rice supply and reported consumption. While they acknowledge that some of the discrepancy could be because of timing issues in sample survey data, the gap is too large to be due to measurement errors alone. They conclude that the figure "indicates possibly significant pilferage."

For India, using data from 1986 to 1987, Howes and Jha (1992) estimated the average ratio of PDS consumption to supply in 18 major states to be 65 percent, ranging from 5 percent in Haryana to 94 percent in Jammu and Kashmir. That is, on an average there was 35 percent diversion. There does not seem to have been much of an improvement since then as similar estimates have been derived by other researchers. For example, Ahluwalia (1993) estimated that in 1986/87, 37 percent of the supply of subsidized rice and 38 percent of the supply of subsidized wheat were illegally diverted. Dutta and Ramaswami (2001) estimated these figures for 1993/94 for the states of Andhra Pradesh and Maharashtra. They found illegal diversions to be of the order of 15 percent for rice in Andhra Pradesh and 30 percent and 19 percent, respectively, for rice and wheat in Maharashtra. A study by Tata Economic Consultancy Services (1998) found illegal diversions to be 31 percent and 36 percent for rice and wheat at the all-India level in the late 1990s. The Planning Commission study (2005) that examined leakages in India after the implementation of the targeted PDS concludes that illegal diversion of rice and wheat at the all-India level in 2003/04 was 37 percent of the total supply of subsidized grain meant for the BPL category.

To get more recent estimates of illegal diversions, we use the National Sample expenditure survey of 2004/05. In that year, the per capita consumption of subsidized food grains was 1.03 kg per month while the per capita supply of subsidized food works out to be 2.27 kg per month. This works out to a leakage of 55 percent of subsidized foodgrains supply. In 1999–2000, these numbers were 1.01 kg and 1.61 kg per month, respectively.[14] These discrepancies are large and suggest a serious problem with diversions.

Table 6 displays the percentage leakages by commodity and according to the subsidy category (POP, BPL, and APL). The aggregate leakage for rice is 40 percent and expectedly diversions are greatest from POP allocations and least for APL allocations.[15] The aggregate leakage for wheat is 73 percent and the diversions are high for all the categories.

14. Because of a change in sample design, the 1999–2000 estimates of per capita consumption of subsidized food could be an overestimate.

15. In comments to the authors, Reetika Khera suggests that grain supplied for the purpose of POP households may be diverted by the states to other households exaggerating the impression of diversion from POP allocations.

TABLE 6. Illegal Diversions as Percentage of Supply, India—2004/05

	Rice	*Wheat*
POP	72	78
BPL	44	70
APL	5	77
Total	40	73

Source: Computations using data on supply of subsidized foodgrains from the Ministry of Consumer Affairs, Food and Public Distribution and data from the Expenditure surveys of the National Sample Survey.

Note: APL = above poverty line; BPL = below poverty line, POP = poorest of the poor.

Excess Costs

All government agencies incur costs in purchase, transport, and distribution of subsidized food. Since this is an activity also done by private agents, it is useful to compare government costs with private costs to ascertain the efficiency of government interventions. In their review of literature about distribution costs, Jha and Srinivasan show that private traders operate at costs lower than those incurred by the government agency in the areas of marketing, storage, trade, and transport despite several controls and restrictions imposed upon them.[16]

In India, the government publishes the "economic cost" of its intervention agency in procuring, transporting, and distributing grain to various stock points. This together with the additional distribution cost to the retail outlets is the government's cost of delivering grain. By comparing it with retail prices of grain, the efficiency of government operations can be evaluated.

Dutta and Ramaswami (2001) used the above methodology to demonstrate that in 1993/94, 27 percent of government budgetary expenditure on food subsidy in the state of AP was wasted by inefficiency of government agencies. The figure for the state of Maharashtra in the same year was 16 percent. A more recent study (Planning Commission, 2005) finds that in the year 2003/04, delivery through the private sector was more efficient in all states except Kerala. The evidence indicates that at the all-India level, the government's food subsidy costs would have been lower by 35 percent if the government costs matched that of the private sector.

In 2004/05, the central government's economic cost of distributing rice and wheat were ₹13.29 and ₹10.19, respectively. To this must be added

16. Jha and Srinivasan (2004) note that the trading costs and wholesale marketing margins of private traders in 2000–01 were about half those of the government agency for wheat and about three quarters for rice.

margins for wholesalers and retailers, and transportation charges at the retail level. We do not have estimates of these costs for 2004–05. A comparison of economic costs with retail prices will therefore give a lower bound to the "excess" costs incurred by the government. The NSS consumption expenditure data for 2004/05 provides information about quantities and expenditures on various items by households. A unit value can be derived from this information. As richer households buy higher quality grain, their unit values are higher. Table 7 displays mean unit values for POP, BPL, and APL cardholding households. Because of large quality variation in rice, prices paid for rice are lowest for POP households and highest for APL households. In wheat, mean prices are about the same between BPL and APL households but are lower for POP households.

TABLE 7. Mean Unit Values for Rice and Wheat in India, 2004/05

Household type	*Price paid for rice (₹/kg)*	*Price paid for wheat (₹/kg)*
POP	9.98	8.58
BPL	10.5	9.34
APL	12.03	9.28

Source: Computations from the Expenditure surveys of the National Sample Survey.

Note: Prices refer to unit values here: APL = above poverty line, BPL = below poverty line, POP = poorest of the poor.

As TPDS grain quality is generally considered to be below average, we take the price paid by BPL cardholding households to be representative for such quality grain.[17] Comparing with the economic costs of the state agencies in 2004/05 (₹13.29 per kg for rice and ₹10.19 for wheat), we obtain the difference as excess cost. The excess cost for rice is ₹2.80 per kg and that for wheat is ₹0.85 per kg.

Direct measures of government costs do not exist for the Philippines. We construct these measures from the NFA's financial statements. Adding the cost of imported rice, operating expenses and interest, we get the total cost as 40,090 million pesos (Table 8). Dividing by the volume of grain distributed (1.57 million metric tons), we get the per unit cost of NFA's rice distribution as PhP 25.5 per kg. The NFA also publishes the market price as PhP 23.56. Hence the excess cost is PhP 1.92 per kg of rice.

17. The data also shows that for both commodities, at least 75 percent of the reported unit values are below the economic cost.

TABLE 8. Excess Cost in the NFA Program, 2006

Volume of rice sold (million metric tons)	1.57
Cost of sales (billion pesos)	31.82
Operating expenses (billion pesos)	3.6
Interest (billion pesos)	4.7
Total cost (billion pesos)	40.12
Per unit acquisition and distribution cost (pesos/kg)	25.48
Market price (pesos/kg)	23.56
Per unit excess cost (pesos/kg)	1.92

Sources: NFA (2006); authors computations.
Note: kg = kilograms.

Measures of Percolation

In this section, we bring together the various components to fit into the conceptual framework outlined in sections three and four. Table 9 summarizes the targeting performance, illegal diversions and excess cost of the food subsidy schemes in India and the Philippines. It is interesting to note that India's TPDS, despite being a targeted program, brings only one-third of the total subsidy to the poor in contrast to the Philippines' universal program that gives them as much as 60 percent of the subsidy. The latter also includes relatively fewer nonpoor among the beneficiaries while incurring lower excess costs that capture the inefficiency of the government-run program vis-à-vis the private sector. However, the food-subsidy programs in both the countries have similar exclusion errors and diversion of subsidized grain supplies to the market.

TABLE 9. Summary of Targeting Performance, Illegal Diversions and Excess Cost

	India	*Philippines*
Exclusion error (percentage of poor)	70	76
Inclusion error (percentage of beneficiaries)	70	48
Share of poor in subsidized grain	33	60
Diversion as percentage of supplies	55	54
Excess cost (as percentage of government cost, rice)	21	8
Excess cost (as percentage of government cost, wheat)	8	–

Source: Our computations from expenditure surveys and official documents as reported in text and Tables 3, 4, 6, 7, and 8.

The last five rows of Table 10 present the components of equation (5) for the Philippines. Note that the total cost figures obtained here are lower than the published food subsidy figures because the latter includes other items such as the cost of maintaining stocks. In the Indian case, the calculations are

TABLE 10. Decomposition of Subsidy Costs in the Philippines, 2006

1	Market price (P/kg)	23.56
2	Value of sales (P billion)	26.61
3	Volume of sales (million tons)	1.57
4	Unit price of sales (P/kg) (item 2/item 3)	16.92
5	Consumer subsidy (P/kg) (item 1–item 4)	6.64
6	Per unit excess cost (from Table 8)	1.92
7	Illegal diversions (million tons) (54 percent of item 3)	0.85
8	Subsidized rice consumed by households (million tons)	0.72
9	Share of poor in subsidized rice (from Figure 1)	0.6
10	Income transfer to poor (item 5 × item 8 × item 9), P billion	2.9
11	Income transfer to nonpoor, P billion	1.9
12	Cost of illegal diversions of rice (item 5 × item 7), P billion	5.6
13	Total excess cost (item 3 × item 6), P billion	3.02
14	Total cost of subsidy, P billion	13.5

Sources: NFA (2006); CEIC Data Company Ltd., authors computations.
Note: P = pesos, kg = kilograms.

a little more cumbersome because of the three layers of subsidy and because of multiple commodities. Tables 11, 12, and 13 lay out the computations and numbers for diversion costs, excess costs and income transfers. The decomposition of subsidy costs into its components is presented in Table 14.[18]

TABLE 11. Diversion Costs, 2004/05—India

	POP	*BPL*	*APL*	*All*
Rice				
Market price (₹/ton)	10,500	10,500	10,500	
Sales price (₹/ton)	3,000	5,650	7,950	
Consumer subsidy (₹/ton)	7,500	4,850	2,550	
Illegal diversions (million tons)	2.3	4.38	0.15	
Cost of illegal diversions of rice (₹ million)	17,250	21,243	382.5	38,875.5
Wheat				
Market price (₹/ton)	9,340	9,340	9,340	
Sales price (₹/ton)	2,000	4,140	6,100	
Consumer subsidy (₹/ton)	7,340	5,200	3,240	
Illegal diversions (million tons)	1.77	5.23	2.47	
Cost of illegal diversions of wheat (₹ million)	13,021.16	27,196	8,002.8	48,219.96
Total cost of illegal diversions				**87,095.46**

Sources: Economic Survey, Government of India; authors' computations.
Note: ₹ = rupees, APL = above poverty line, BPL = below poverty line, POP = poorest of the poor.

18. It is well known that NSS aggregate population counts are lower than the Census total. In the estimates displayed here, the estimates of consumption from NSS are blown up to the Census numbers.

TABLE 12. Excess Costs in India, 2004/05

	Rice	*Wheat*	*All*
Economic cost (₹/ton)	13,296	10,190	
Market price (₹/ton)	10,500	9,340	
Per unit excess Cost (₹/ton)	2,796	850	
Quantity sold (million tons)	16.46	12.89	
Total excess cost, ₹ million	**46,033.34**	**10,956.5**	**56,989.84**

Sources: Economic Survey, Government of India; authors' computations.
Note: ₹ = rupees.

TABLE 13. Income Transfers, 2004/05—India

	POP	*BPL*	*APL*	*All*
Rice				
Market price (₹/ton)	10,500	10,500	10,500	
Sales price (₹/ton)	3,000	5,650	7,950	
Consumer subsidy (₹/ton)	7,500	4,850	2,550	
Consumption of subsidized rice (million tons)	0.90	5.65	3.15	
Share of poor	0.47	0.34	0.21	
Income transfer to poor (₹ million)	3,193.30	9,415.55	1,646.83	14,255.68
Income transfer to nonpoor (₹ million)	3,549.20	17,986.95	6,385.67	27,921.82
Wheat				
Market price (₹/ton)	9,340	9,340	9,340	
Sales price (₹/ton)	2,000	4,140	6,100	
Consumer subsidy (₹/ton)	7,340	5,200	3,240	
Consumption of subsidized wheat (million tons)	0.50	2.19	0.73	
Share of poor	0.53	0.41	0.22	
income transfer to poor (₹ million)	1,922.26	4,663.72	509.89	7,095.87
Income transfer to nonpoor (₹ million)	1,718.38	6,724.28	1,855.31	10,297.97
Total income transfer to poor (₹ million)				**21,351.55**
Total income transfer to nonpoor (₹ million)				**38,219.79**

Sources: Economic Survey, Government of India; authors' computations.
Note: ₹ = rupees, APL = above poverty line, BPL = below poverty line, POP = poorest of the poor.

TABLE 14. Decomposition of Subsidy Costs (India, 2004/05)

Income transfer to poor (₹ million)	21,352
Income transfer to nonpoor (₹ million)	38,220
Illegal diversion cost (₹ million)	87,095
Excess cost (₹ million)	56,990
Total cost of subsidy (₹ million)	203,657

Source: Tables 11–13.
Note: ₹ = rupees.

Table 15 displays for India and the Philippines the percolation metric from a unit of public spending on the poor. The share of subsidy going to the poor is 11 percent and 21 percent, respectively, in India and the Philippines. Multiplied by the participation rate of the poor, the percolation indices are 0.05 or less.

Table 15. Percolation to the Poor

	India	*Philippines*
Total subsidy	₹ 204 billion	P 13.5 billion
Income subsidy to the poor	₹ 21 billion	P 2.9 billion
s—share of subsidy received by poor	0.105	0.214
Participation rate (percentage of the poor)	30	24.5
Percolation index	0.03	0.05

Source: Authors' computations.
Note: ₹ = rupees, P = pesos.

The pie charts in Figures 4 and 5 graphically display how the subsidy is spent on various components. Interestingly, the illegal diversion is identical, and the excess costs are similar between the two programs. While the poor's participation rates are similar as well, there is a striking difference between the NFA and the TPDS on the share of subsidy going to the poor. We put forth three reasons to explain why the share in TPDS is only half of that in NFA although the NFA is a universal program. All the three reasons reflect indirect targeting. First, NFA rice is, by law, clearly labeled as such. This may attach a stigma to the rich who often buy it from supermarkets or malls where it is retailed. Second, anecdotal evidence suggests that an informal quota of 5 kg per head is implemented in practice though there is no limit on the number of times one may queue up. This would add significantly to the opportunity cost of the nonpoor. Third, the NFA has procedures for renewing retailers' accreditation, which may not be granted if consumers complain enough (Mehta and Jha, 2010).

Figures 4 and 5 show that even if inclusion errors were minimized to zero, the share of the poor in the subsidy would rise at most to 35 percent in Philippines and to 29 percent in India. This means that the percolation metric would rise to about 0.09 in both countries. While this would be a significant rise over the existing situation, the percolation metric would still be much closer to zero than to one. This shows that improvements in the percolation index by giving positive utility weights to incomes just above the poverty line would not materially alter the magnitude of percolation.

FIGURE 4. Decomposition of Subsidy–Philippines

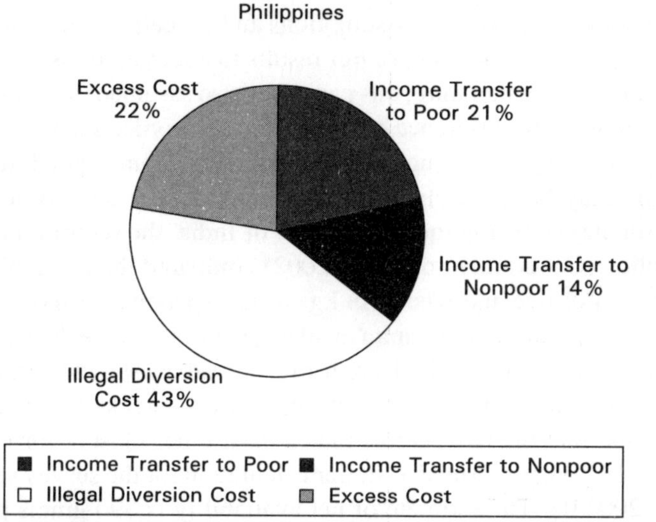

Source: Table 10.

FIGURE 5. Decomposition of Subsidy–India

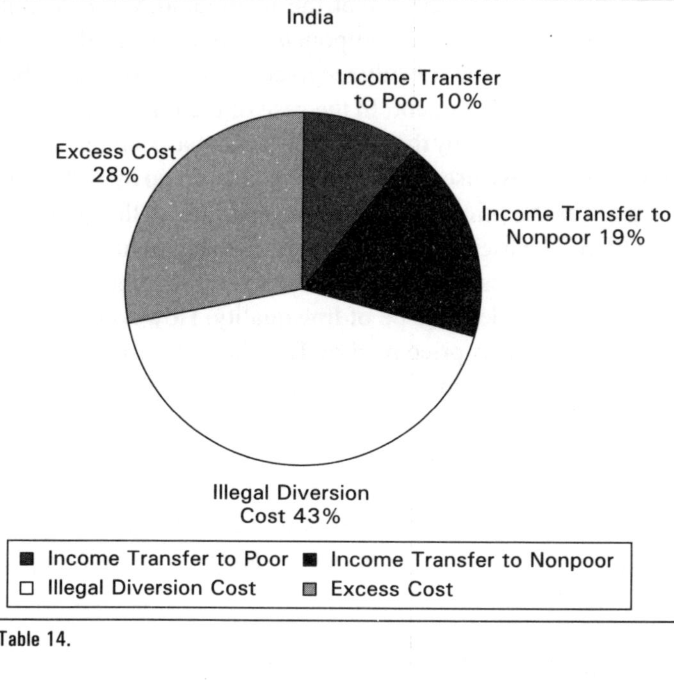

Source: Table 14.

Assumptions and Robustness Checks

In this section, we record the assumptions and caveats to our methodology. We also report the robustness of our results to alternative assumptions.

Like much of the literature, this paper has also assumed that leakage (illegal diversions) is the difference between official recorded sales of subsidized grain and consumption of such grain as estimated from expenditure survey data. This may be misleading if expenditure survey data systematically underestimates consumption. In the case of India, the report on long-term grain policy (Government of India, 2002) compared the per capita annual consumption of rice and wheat (in kg) from expenditure survey data with the official estimates of per capita availability of these cereals (i.e., production minus different uses which are for seed, feed, waste, industrial use, net exports, and change in public—but not private—stocks and waste). The report found that the two figures match each other closely for the period 1991–2001.[19] We performed a similar calculation for the survey year of our interest, 2004/05. The estimate of net availability is 390 grams per capita per day while the estimate from the NSS consumption survey is 386 grams per capita per day. There is, therefore, little reason to fear any substantial bias from the method of computing leakage.

A second possible concern is that the methodology for decomposing public expenditures into various components (Figure 4) needs to identify a market price to compute the subsidy received by households and the excess costs incurred by the government. In the case of the Philippines, we rely on market price data published by the NFA. We do not have similar authoritative data for India. Instead, we used the average of prices paid by BPL households for market grain as the price that is most comparable to the quality of grain sold in the TPDS. The logic of this procedure is that purchased grain quality (and hence prices) varies directly with household income and that TPDS grain is generally considered to be of low quality. However, the identification of the market price as price paid by BPL households is arbitrary.

From equation (4), it can be seen that at the margin, a higher market price will reduce excess cost and increase the income transfer to households. However, this also results in a greater share of illegal diversions in public expenditures. Overall, the amount of income transfer goes up by $q \Delta p$ and total program waste (sum of leakages and excess cost) falls by the same amount. Suppose the market price of rice is identified as the 75th percentile

19. See Annexure II of the report.

of the distribution of unit values in the survey data. Suppose also a similar assumption for wheat. Under these assumptions, the market price of rice is ₹12.66 per kg while that of wheat is ₹10 per kg. These are improbably high values for TPDS quality grain—nonetheless, it is worth seeing how the estimates change. Given the estimates of q, these market prices imply that the total income transfer goes up by ₹23,209 million while the program waste falls by the same amount. In terms of the pie chart in Figure 4, the income transfer to households rises from 29 percent to 40 percent. While this is a substantial rise, program waste still accounts for 60 percent of all public expenditures. Within this category, the share of excess costs falls while that of leakages increases.

We also considered another alternative procedure which again produces upper bounds on the amount of public expenditures transferred to households. The prices that households pay vary both spatially and temporally. Consequently, the subsidy received by households would also vary similarly. To take this into account, we consider the subsidy received by households as the product of the quantity purchased from TPDS and the difference between the prices paid for market and subsidized grain. This would be accurate as long as the market grain is of the same quality as subsidized grain. However, if households buy grain of varying quality, this procedure will overestimate the subsidy received especially by richer households.

A complication in applying this procedure is that about 4.4 percent of households buy grain exclusively from the PDS. For these households, a market price must be imputed to calculate the subsidy received by them. For this reason, we follow Deaton (1997) in regressing unit values on household expenditure, household size, and village fixed effects (or urban block in urban areas). From this equation, we obtain predicted unit values for all households in the sample. These are used as the relevant market prices to compute the subsidy income received by households. By this procedure, the aggregate subsidy received by households rises to ₹68,640 million (from ₹59,572 million) in the benchmark estimates reported in Table 14.[20] The proportion of public expenditures received by households rises to 34 percent from 29 percent reported in Figure 4. In conclusion, neither of these procedures, that are engineered to estimate an upper bound on the income subsidies to households, results in estimates that warrant revision of the qualitative story told by the benchmark estimates of Table 14 and Figure 4.

20. Once again, as in the benchmark case, the numbers are adjusted upward to take account of the downward bias in NSS population aggregates (relative to the Census).

It should be understood, however, that our estimates of percolation apply only to incremental changes in public expenditure. The methodology implicitly assumes that the average participation rates and the share of income transfer going to the poor (from equation [6]) are valid for the increment in public expenditures. Lanjouw and Ravallion (1999) point out that the average participation rate of the poor in antipoverty programs may in fact be a poor guide to their marginal participation rate induced by additional public spending. In particular, if there is early capture of these programs by the nonpoor, then it could well be that the average participation rate of the poor severely underestimates their marginal rate. Lanjouw and Ravallion use cross-sectional variation in participation rates to illustrate their thesis for antipoverty programs in India, such as public works and asset transfers. Notably, however, they find that average participation rates do not differ much from marginal participation rates for the public distribution system. Even otherwise, the relevance of their argument is limited in a world where most of the public expenditure is absorbed by illegal diversion and excess costs.

It should be clear, however, that this paper's methodology is entirely partial equilibrium in nature. This applies especially to our computation of subsidy and excess cost. It could be argued that the subsidy received by households should be evaluated with respect to a counter-factual: what would market prices be if there were to be either no intervention or an intervention of a different form (say, food stamps or cash transfers). Such a general equilibrium analysis is not attempted here.

Policy Options

Neither India nor the Philippines score well on the percolation index. Participation rates are low and households, whether poor or not, do not receive most of the expenditures of the food subsidy. The in-kind subsidies offered in India and Philippines pose particular challenges with respect to illegal arbitrage and fraud and with respect to the excess costs of state agencies. It is possible that these problems are generic to in-kind transfer programs. For instance, illegal diversions have also been reported for Indonesia which too has an in-kind food subsidy program. Olken (2006) estimates that minimum leakages in Indonesia are of the order of 18 percent of the supply of subsidized rice. More realistic assumptions lead to estimates of around 30 percent.

The impact of food subsidies on the poor can be increased either by increasing the participation rate or by enhancing the fraction of subsidy going to the poor or a combination of the two. Policies aimed at the latter will save resources that could be used to increase the participation rate.

The scope for this can be seen clearly in the Indian case. Figures 6 and 7 plot the scatter and the line of best fit between the participation rates of the poor and the nonpoor across Indian states in 1999/2000 and in 2004/05. The correlation is visibly very strong and the R^2 is in excess of 0.9 in both years. Quite clearly, the costs of participation for the poor are positively correlated with the costs of participation of the nonpoor. An ideal targeting system would be one where the costs are high for the nonpoor and low for the poor. It is clear that such a system is not in place, and if such a system cannot be devised, then the food subsidy system would have to be near-universal if the poor are to participate in large numbers.

The difficulties of devising a targeting system that approximates the ideal are formidable. Most of India's work force is either self-employed as farmers, traders, vendors, craftsmen or they are wageworkers in the informal sector of trade and manufacturing. Such employment is characterized by the absence of formal contracts, salary records, and tax payments. Means-testing as is practiced in developed countries is impossible. Identification of poverty status depends on proxy indicators of land ownership, habitation, type of housing, and social characteristics. It cannot be expected that these would perfectly correlate with poverty status defined by the official poverty line.

Given that the official poverty line in India measures bare subsistence, can any set of proxy indicators finely differentiate between households above the poverty line and those below it?[21] Table 16 shows that in the universe of households that are deemed to be eligible for either POP or BPL benefits, 32 percent in rural and 49 percent in urban areas have per capita consumption expenditures below the poverty line. In other words, 68 percent of POP/BPL households in the rural sector and 51 percent of such households in urban areas are not poor (by the official poverty line). This suggests that inclusion errors are serious in the distribution of POP and BPL eligibility cards. However, the table also shows that 70 percent of POP/BPL rural households

21. Jalan and Murgai (2006) show that the proxy indicators used in India to arrive at a census of poor households in 2002 were unable to differentiate between extremely poor and not-so-poor households. Enlarging the set of proxy indicators to include other household characteristics did not help much possibly because of the bunching of households around the poverty line.

FIGURE 6. Scatter between Participation Rates of Poor and That of Nonpoor in the Public Distribution System, India, 1999/2000

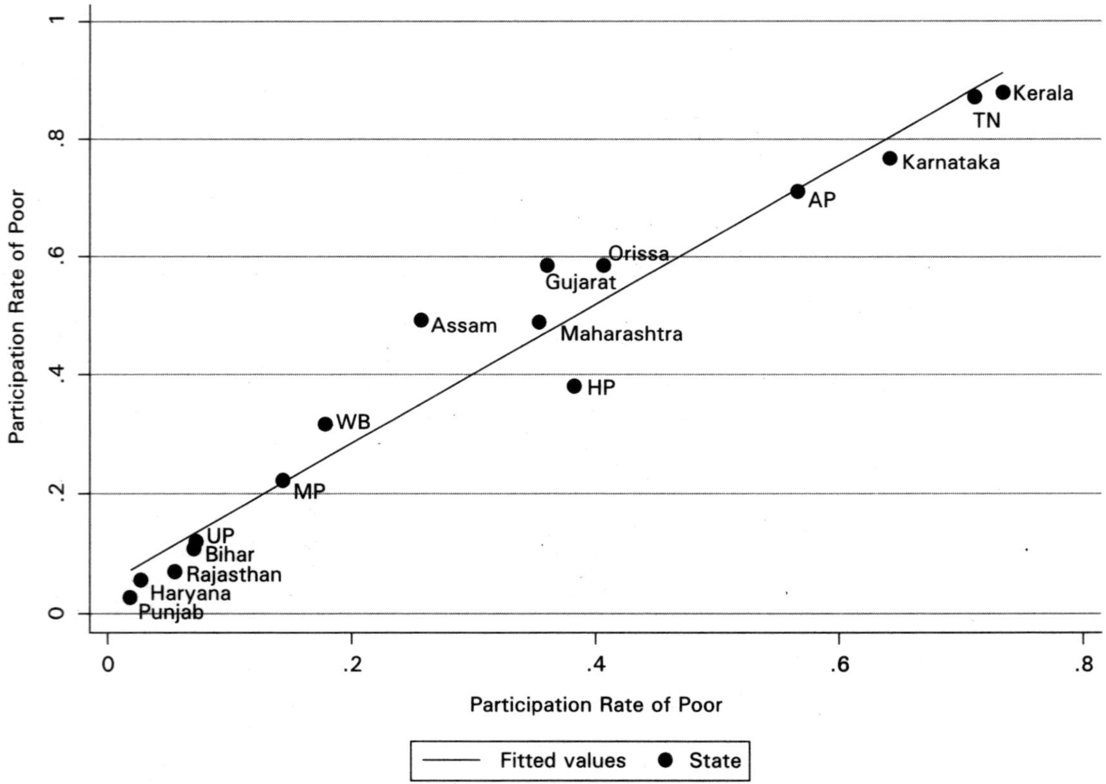

Source: Computed from the Expenditure Survey of the National Sample Survey.

FIGURE 7. Scatter between Participation Rates of Poor and That of Nonpoor in the Public Distribution System, 2004/05

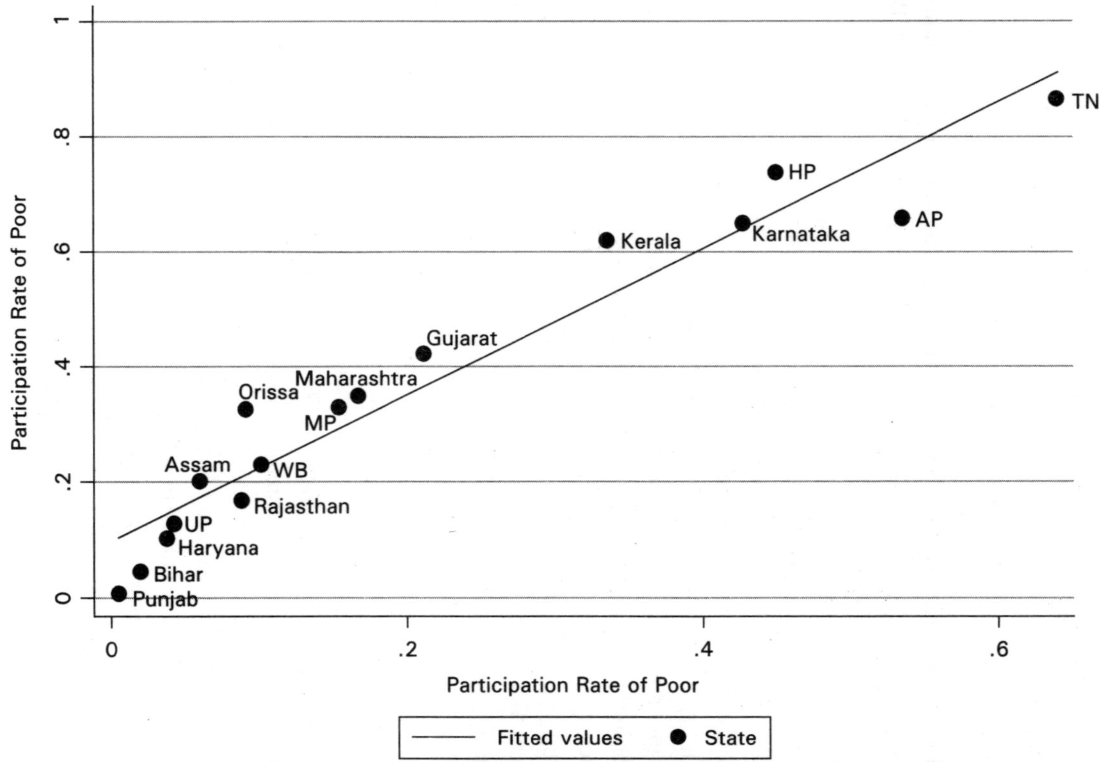

Source: Computed from the Expenditure Survey of the National Sample Survey.

TABLE 16. Proportion of Population with POP and BPL Eligibility in the Following Expenditure Categories, India, 2004–05

Category	*Rural*	*Urban*
Expenditures below poverty line	32	49
Expenditures below 1.5 times the poverty line	70	78
Expenditures below twice the poverty line	87	90

Source: Computed from the Expenditure Survey of the National Sample Survey.

(and 78 percent in urban) are those with per capita consumption expenditures below 1.5 times the poverty line. Many of the nonpoor with POP or BPL cards are those with per capita expenditures just above the poverty line.

The problem with targeting in India is therefore not so much that grossly ineligible households have been counted in, but that many deserving households have been left out. Anecdotes from officials suggest that one reason for this is that the proxy indicators throw up estimates of eligible households far in excess of the estimates of BPL households. State governments are then under pressure to trim the list of eligible households to match BPL estimates and this bureaucratic process leads to exclusion errors. Indeed, with 78 percent of the rural population and 61 percent of the urban population having per capita expenditures less than twice the poverty line, a targeting system that insists on matching the list of subsidy-eligible households with the estimates of poverty will likely have large exclusion errors.

The dilemma is that while a move toward enlarging the number of eligible households will increase participation rates and hence percolation, it would also also increase inclusion errors and perhaps decrease the share of the poor in food subsidy expenditures. Fortunately, however, the waste in the food subsidy systems of both countries (of 65 percent and above) provides a buffer by which coverage can be stepped up substantially without commensurate increase in public expenditures.

It is this reason why efficiency of subsidy delivery is the key to food subsidy reform. This is a much debated issue in India. One option is to continue with in-kind transfers and to reform the public distribution system in the direction of greater efficiency. Advocates of such policies point to the success of states like Chhattisgarh and Tamil Nadu in achieving efficiencies in the public distribution system through a mix of policies (enlarged coverage, nationalizing the ration shops) and information technology systems in policing the movement of grain through the distribution network. Such policies require sustained political and bureaucratic commitment to the supervision of the supply chain and it is not clear that this is replicable by other states. Another option is to insist on reliable authentication of retail transactions with the use of smart cards and/or biometric identification. Such authentication would make it difficult to divert grain from genuine beneficiaries. These reforms, however, would not directly attack the problem of excess costs.

On the other side of the debate are those who propose that in-kind transfers be replaced with cash transfers. Not only would cash transfers get rid of illegal diversions and excess costs, but they would also remove the distortionary impacts of in-kind transfers on consumption and production. Cash transfers are often criticized for being mere income transfer programs. In-kind transfers are regarded as more appropriate if the objective is to meet specific targets of food intake. It can be debated whether paternalism should be the guiding principle or whether consumer sovereignty ought to be respected. However, transfers of any kind are fungible in the hands of the beneficiary, and it is questionable whether in-kind or cash transfers can achieve desired norms of nutritional intake (Jensen and Miller, 2011). The more serious critiques against cash transfers are that it calls for a reliable infrastructure of payment systems (whether through post offices, banks, or other means) and that it must be price-indexed if their value is not to erode over time.[22]

This contest between an untested system (cash transfers) that holds promise to some and fears for others and a tried system (in-kind transfers) that has failed in all but a handful of regions calls for a policy framework that allows for experimentation, learning, and adjustment. Such a policy framework runs against the bureaucratic impulse to govern through uniform formulaic mechanisms.

More fundamentally, there are formidable political economy obstacles. Government procurement of grain is a powerful symbol of its commitment to rice and wheat growers. Direct income transfers to consumers also support foodgrain demand, but it lacks the visible assurance to counter the political mobilization that is possible with a demand for support prices. For this reason, the central government cannot be seen as dismantling in-kind transfers. Reforms are more likely if individual states (especially those that import food from other parts of the country) have the freedom to tailor the food subsidy system to their needs. At the state level, the political economy is straightforward. The government will have to weigh the political gains of an effective food subsidy regime against the lobby that derives substantial rents from the distribution of subsidized grain. The demonstration power of a single success will make it harder for state governments to resist success.

22. See also Kotwal et al. (2011) for a discussion of the critiques of cash transfer. The paper also lays out a scheme for the design and implementation of cash transfers in India.

Comments and Discussion

Surjit Bhalla
Oxus Investments Pvt. Ltd.

Though we have not conferred, my comments are very similar to Rinku's and I am in substantial agreement with most of what she said. The paper has some important findings and my comments will be more in the nature of what can be done more. There is no problem that I see in the percolation index. I do not know whether that was a mandate that you got from NCAER, etc., but this is what I would encourage. Let me illustrate from that.

First, and most important, what is missing in the paper is an emphasis on the political economy of the system and in schemes like the PDS. As mentioned in the paper, Kirit Parikh (1994) documented the nonworkings of the PDS. Despite several additional investigations, the basic conclusion has not changed. PDS has not worked; so the question is: Why is the political economy background of the PDS not evaluated in detail? Let me try and suggest what I mean by the "political economy of PDS" and what can be done.

First and foremost, why do we need, in order to have a food subsidy program, something called the FCI? Why is that necessary? Is there any country in the world that has a similar program? It does not seem like Philippines has anything even comparable. In India, the government sets the procurement prices, collects the foodgrains via the FCI, stores the food, and then distributes the foodgrains on the basis of whether you are poor or not and does so via only government-authorized ration shops. The idea is that if you want to distribute food to the poor at a subsidized price, why is the simple solution not adopted in India—any food shop can distribute the food and the recipients have an allocation, e.g., food stamps? This does not involve the FCI, nor procurement policies, nor government ration shops.

Second, as far as the political economy is concerned and the historical record, the paper has no mention of the two countries that have practiced food subsidies for a very, very long time. US instituted the food stamp program in the mid-1960s and Sri Lanka in 1979. In India, the problem with the policymakers as well as the policies is that we believe we are sui generis, we are unique and that nobody else has faced this problem. But they have and they do. How has the performance of these two countries been with the food stamp program? What kind of leakages they have in their system versus the leakages in our own system? A political economy treatment will involve such evaluations.

The question remains: what do we do now? Several possibilities, especially if we want to do something constructive. To prevent leakages (via purchase of food stamps or even the present system), why not distribute only low-quality foodgrains? An alternative policy is cash transfers. What are the objections to cash transfers? One objection is that cash transfer recipients may not have bank accounts. Here is an alternative which should be considered but which the political economy may not allow. Let panchayats be responsible for the distribution of the food or the money or whatever else. If the panchayat leaders are leaky, the people, the media, the nongovernmental organizations (NGOs), the World Bank, Asian Development Bank (ADB), etc., can expose them. A third option, but not political economy correct option, is to target the entire population in the 33 percent poorest districts of India. Why has this option not been considered? What is the problem with this recommendation?

Last, the unique identification (UID) program. Look at the political economy of this extremely worthwhile program. Again, the lobbyists have begun to discredit this program. They claim (and many of these anti-UID lobbyists are part of the government, part of the intellectual National Advisory Council [NAC] elite), that such a program invades privacy, cannot work in a country like India, etc. We are argumentative Indians; we are Indians who argue without any knowledge of facts because facts are not relevant. What I would suggest to the authors is that they take the arguments against UID, or cash transfers, of experts like Mr Jean Dreze, and Mr N.C. Saxena. If the facts support the views of these experts, it will embellish the paper; if the facts reject, it will embellish the paper.

Rinku Murgai

The World Bank

This is an interesting paper which raises some important empirical and policy issues. The paper looks at the extent to which expendi-tures on food subsidies reach the poor in India and the Philippines. The authors modify a commonly used measure of targeting, the share of subsidy that accrues to the poor, to propose a percolation index which combines the share that accrues to the poor with the percentage of the poor that receive the subsidy.

This is meant, in their words, "to capture the impact of a marginal expansion of the program on the poor."[23] In my view the main contribution of the paper is that it takes account of both targeting errors—inclusion and exclusion—and nontargeting leakages in estimating the share of expenditures reaching the poor. Both sources of funds diversion (away from the poor) are well acknowledged in the literature, but very rarely do papers combine both. This paper shows how important it is to look at the nontargeting leakages combined with the targeting leakages in order to understand the extent to which government expenditures are reaching the poor. This I think is the primary contribution.

In order to estimate the share that is reaching the poor, the authors decompose the subsidy into four components. One is the excess cost of State agencies relative to presumably what would have been cost of distributions by a more efficient private sector, illegal diversion and the rest, which is either subsidy transferred to the nonpoor or subsidy transferred to the poor. Their central finding is that in both countries participation rates by the poor are low and households, poor or not, do not receive a large share of the subsides. The findings suggesting that excess cost plus illegal diversion absorb 65 percent of expenditures in the Philippines and 71 percent in India are indeed startling.

I have three main comments on the method. First, what does the percolation index measure? It is estimated as product of the share of the subsidy received by the poor with the share of the poor who receive the subsidy. When the index takes a value of one, most of the expenditures reach the poor and they reach a large percentage of the poor. If the index takes the value zero, the poor receive very little benefit from the program. This would only capture the percolation of a marginal expansion of expenditures on the poor, if the share of the poor in the incremental subsidy would be the same as what is observed on average, i.e., the marginal share is equal to the average share. This, a priori, is not a harmless assumption. For example, a paper by Peter Lanjouw and Martin Ravallion that estimates the marginal incidence of selected antipoverty programs in India concludes that average participation rates are not a reliable guide to how the scaling up of a program would affect participation by the poor. They use interregional or interstate differences in program scale to infer how the composition of participation between poor and nonpoor varies across, and thereby estimate marginal incidence. It would

23. I have quibbles with both terms "impact" and "marginal." I prefer to not use the term impact because this is not the traditional sense in which we think of impact on welfare or impact on final outcomes of interest. But it is still a useful summary measure of how much of the expenditures are reaching the poor.

be worth following a similar approach in this case as well. If an assumption of homogeneity is not rejected, I would have more confidence in the results. This is in the nature of quibbles. Even so, the percolation index proposed in this paper is still useful as a summary measure of the average percolation of expenditures to the poor.

The second main comment I have is on the calculation of excess costs. Excess costs are measured as the total supply of grain distributed by the governments (offtake) valued at the difference between the economic cost and the market price paid by low-income households. The implicit assumption that is being made is that the observed market price captures well the counter factual: what it would have cost the government to procure and distribute the grain if it were as efficient as the private sector. I'd like to note a few issues related to this.

First, there is bound to be excess cost in government acquisition and distribution given the way the system works in India. In some areas (not all), the system involves purchase of grain by the government, carting it out of rural areas and then carting it back to rural areas. Some dead weight loss is to be expected and unless the system is reformed to have a much more decentralized scheme for procurement and distribution, it is hard to avoid it entirely.

Second, and more directly pertinent to the estimation in this paper, the question I have is whether the average unit value paid by BPL households (and presumably what is being measured is for non-PDS grain only) *nation-wide* is an accurate measure of the price that FCI or some other State agency would pay if they were more efficient. It is not obvious to me that that is necessarily the case. There will be general equilibrium effects of FCI intervention on retail prices and it is worth thinking about whether FCI, being such an important player in the market, is propping up or depressing retail prices. What we need to get at is the counter factual: What would be the price if FCI operated more efficiently? Second, on the same topic, as a simple check, it may be worth comparing the unit values of urban and rural areas within each state where one could think of the difference between the urban and rural retail price as a measure of the private sector's cost of moving grain. How does that compare to the reported estimates of economic costs? Also in case data is available, it would be interesting to do this calculation state-wise and separately for urban and rural areas. I am pretty sure you will still get very high levels of excess costs, but some fine-tuning of the methodology would inspire more confidence.

My final point main comment is: Where do we go from here? I agree with the view that given the contest between an untested system (cash transfers) and a tried system (in-kind transfers) that has failed in all but a handful of regions, there is a need for a policy framework that allows for experimentation, learning, and adjustment. I completely agree with that. But I would like the authors to try to go a bit further with the data at hand.

What do we know? One, several studies show that targeting can work reasonably well at the tails. You can do a reasonably good job excluding, say, the richest 25 percent, and with some (less) confidence include the bottom 15 percent. The middle is a gray area and it is bound to have errors in design and, not to mention, in practice. It is very, very difficult to design a targeting system that is going to do a reasonable job in the middle of the distribution. So we are going to live with targeting errors. State-level experiences which the authors cite suggest that to significantly enhance participation by the poor, a near-universal program (I am not saying whether it is cash or in kind), is the way to go. What about the share of subsidy going to the poor? The authors claim that "enlarging the number of eligible households will increase participation rates" but "perhaps decrease the share of the poor in food subsidy expenditures." I think this question needs to be examined more closely. Let us focus on a few states where the number of eligible households has been increased by states like Tamil Nadu, Himachal Pradesh, Andhra Pradesh, and Chhattisgarh to see whether this is the case or not. Finally, there is the question of what is it that we are trying to achieve and here the authors do not actually lay out the arguments. I would like to see them laid out. Is it about income support to the poor? If yes, there is a case for cash. Is it about food security, and if so, is that possible with cash transfers? Is it about price stability? If yes, what does that imply about how offtake is to be managed if transfers are to be made in cash? These are some questions. At least the questions should be laid out and some thinking on how to evaluate those options, I think, would be desirable. Thanks.

General Discussion

Shubhashish Gangopadhyay (session chair) opened the discussion by reiterating the big question about the policy implications of the paper's results. He also had a query about the paper's finding that the poor who have APL cards had a lower participation rate than the poor with BPL cards, and whether this had something to do with the NSS data.

Bharat Ramaswami noted that the paper was written to provide evidence to inform any policy debate, evidence that was badly missing. For example, much of the debate in India on universal versus targeted transfers or cash versus in-kind transfers made no reference to the efficiency of the PDS. This debate did not recognize that any reasonably comprehensive program would have to deal with large targeting errors. The paper's finding that 65 to 70 percent of resources were wasted even before targeting errors suggested the massive room available to increase participation rates by adopting more generous eligibility criteria and yet becoming more efficient if the pre-targeting waste could be reduced.

The question of which was more efficient—cash or kind—was a large topic and taking it on would have reduced the focus of the paper on measuring the wastage in the existing system. He referred participants to a May 2011 *Economic and Political Weekly* (EPW) paper by Kotwal, Murugkar, and Ramaswami that dealt with this topic in a much greater detail.

Referring to Rinku Murgai's comment on the marginal equaling the average, he agreed with her. But he noted that the Lanjouw paper's treatment of marginal was very different from the way this paper was considering marginal to be very, very small changes. Looking at regional variation as done in the Lanjouw paper need not be marginal at all. He agreed that getting to the counterfactual on what market prices to compare for estimating excess costs was not easy. The paper clearly had partial equilibrium estimates. One would have to think of the counterfactual first in the absence of any intervention at all, and then a counter factual in the absence of FCI intervention, since the two could be different. These problems would have to be dealt with even with a computable general equilibrium (CGE) approach. Ramaswami felt that most often the quantum of the actual subsidy was slender, so doing a general equilibrium analysis may not change it that much.

Shikha Jha responded to Surjit Bhalla's question on whether the Philippines had an agency like the FCI by noting that the NFA was similar to the FCI. The only difference was that it imported grain and had the monopoly to do so. The NFA operated a buffer-stock scheme and sold rice at 10 percent lower than market price to consumers and about 10 percent higher than market price to producers. It held about 30 days of stock for price stabilization and about 15 days of stock for emergency use. It procured less than 2 percent of its total requirement domestically because Philippines was import dependent for rice.

On the Chair's query about lower participation by the poor who had APL cards, Shikha Jha responded by noting that this may have to do with the higher prices that APL have to pay. Bharat Ramaswami noted that APL allocations also depended more on the discretion of the central government and the availability of grain.

Indira Rajaraman asked the question of where we go from here given the important findings of the paper. The PDS is not the only food security program in India. There are several price support schemes; there is MGNREGA, which could be seen as a food security scheme that people self-select into by their willingness to do manual work, the mid-day meal scheme, and the Integrated Child Development Services (ICDS). Her plea was to consider these schemes together and then explore what could be done to rationalize them. She felt that because it is self-targeting, MGNREGA could be the nucleus around which India could build food security. Then we would need to see who is vulnerable but cannot self-select because they are either too old, disabled or geographically scattered, and deal with these special situations. She also made the point that if the FCI is to be liquidated, there will have to be a strategy that takes in account the prime land that it owns in many cities and how the proceeds from that land can be helpful in facilitating any exit scheme.

Pradeep Mehta emphasized that PDS and MGNREGA were part of a bigger process of patronage at the grassroots level and this would make finding political solutions extremely difficult. He felt that there was an implicit consensus among political parties not to interfere with the PDS.

Dilip Mookherjee asked about the market prices the paper had used to assess excess distribution costs. Food was procured in one rural area and then moved around the country, and presumably there was a lot of regional variation in prices. He also raised the issue, which he did not see discussed much in the Indian context, of the impact of the PDS on welfare dependence, on mobility, and on the incentives of the poor to escape poverty. How does India's transfer system, somewhere in between unconditional cash transfers and in-kind transfers, impact the incentives for upward mobility? He also noted that though the participation rates looked quite different between APL and BPL, the income transfer estimate was much larger for the nonpoor than for the poor.

T.N. Srinivasan argued that India's many food-related transfer schemes had different objectives. To take all these into account, it was important to do what the authors had done some years back in looking at foodgrain prices in India, i.e., to have a well-specified general equilibrium model that focused also on trade policy issues that could be important for food distribution. He urged for a more detailed general equilibrium analysis that included foodgrain markets.

Govinda Rao noted that in the Philippines food was being imported at international market prices and then distributed locally, whereas in India there were significant additional distortions and related costs that crept in because the government was setting the domestic procurement price, including externalities such as environmental costs. He was also unsure how the paper adjusted for the underestimation of consumption that the NSS data is typically blamed for, unless the paper assumed that the NSS consumer expenditure data is correct.

Ulrich Bartch drew attention to the large economic costs of the PDS because of the many distortions it creates in agriculture. The paper's startling findings would look even more startling if we took into account the costs, e.g., of the distorted supply response and the decline in agricultural productivity because farmers are producing rice in areas not suited for rice, as a result of the government's minimum support price for procuring PDS grain.

Ashok Lahiri wondered why grain could not be procured closer to consumption centers through standard procurement rules that all governments follow. The tenders could, of course, come from all over. Why must the government procure grain only where it is produced and then take the trouble of getting the FCI to transport it to consumption centers? He also wondered why the responsibility for targeting and distribution could not be given to the states, with the Center simply giving them the money.

Bharat Ramaswami responded to the question of why income transfers to the nonpoor were larger while their participation rate was lower. He explained that in their paper the poor were defined using the official poverty line, but the participation rates came from separate data on those who had BPL or APL cards. Since there were many BPL cardholders who were not poor, a lot of the income transfer could be happening to them. To the question of which market price to use for assessing the excess costs of PDS, he felt there was no single, clean answer. In wheat where there was not much quality variation, the issue of which price to use was less important, but in rice it was important. On the question about NSS consumption expenditure data, he noted that people working on the issue have suggested that if all food grains are taken together, the NSS matches the data on availability coming from the supply side.

Surjit Bhalla noted that the estimated cost of illegal diversion in both the Philippines and India was identical at 43 percent. The estimated excess cost of public as compared to private distribution was also broadly similar, at 22 percent in the Philippines and 28 percent in India. But the big difference came in the income transfer to the poor, 10 percent in India and 21 percent in the Philippines, and in the income transfer to the nonpoor, 19 percent in India and 14 percent in the Philippines. The Indian percolation index was approximately half to 60 percent of what the percolation index was in the Philippines.

Rajesh Chadha pointed to the many hidden costs in the public procurement and distribution system in India. For example, misplaced regulation in rice milling in Punjab state thwarted competition among millers, as a recent NCAER survey showed. Millers are required to sell 75 percent of milled non-basmati rice at a predetermined price as a levy to FCI. The only alternative is to mill paddy allotted to them by FCI at fixed rates of milling. Millers respond by mixing in the poorest quality rice or rice fragments into what they mill.

Pratap Bhanu Mehta wanted to ask what the impact of the diversion of grain and its reselling was on market prices. Given the large quantities that the paper estimated are diverted, and then presumably resold, what do we know about the impact of this diversion on market pricing?

Anil Sharma observed that it may be too early to write the obituary of targeting in India since India had mostly run a universal PDS for 35 years and started targeting only some 10 or 11 years ago. The unique biometric id *Aadhaar* could help address some of the weaknesses such as duplication and bogus cards.

Pradeep Mehta wanted to remind everyone that Prime Minister Rajiv Gandhi had said many years ago that only 15 paisa of every rupee the government spent on the poor actually reached the poor, so the estimates in his paper should not have come as a surprise. Disagreeing with Surjit Bhalla, he felt it would be suicidal to hand over the PDS to panchayats. That would only drive the corruption down lower.

Surjit Bhalla replied that continuing to oppose devolution to panchayats missed out on the changing role of technology, civil society, and the media. These changes would increasingly make it very difficult for even panchayats to do the wrong thing and get away with it.

Indira Rajaraman noted that if we accept MGNREGA as the core of a future food security program and add to that an additional program for those who are vulnerable and cannot self-select, then panchayats are especially good at identifying the vulnerable through the Gram Sabha.

Pradeep Mehta clarified that handing over responsibility for regulating PDS shops to panchayats would be more effective than the current practice of asking the district supply officer to do it. But it would not solve the problem. Though it had not come up, he noted that the margin on food grains distributed through the PDS system was so low that unless the owner cheated, he could not make any money.

Shubhashis Gangopadhyay ended the session with his comment that he found it strange that they had been talking about targeting when almost in the same breath others are talking about the legislation that treated food as a basic right. He hoped that this fundamental issue could be discussed in the Roundtable on Subsidies scheduled for the following day at the IPF.

References

Ahluwalia, Deepak. 1993. "Public Distribution of Food in India—Coverage, Targeting, and Leakages," *Food Policy*, 18 (1; February): 33–54.

Besley, Timothy and Ravi Kanbur. 1993. "The Principles of Targeting," in M. Lipton and J. Van der Gaag (eds), *Including the Poor*, pp. 67–90. Washington, D.C.: The World Bank.

Coady, David P. 2002. "Designing and Evaluating Social Safety Nets: Theory, Evidence and Policy Conclusions," FCND Discussion Paper no. 172, International Food Policy Research Institute, Washington, D.C.

———. 2003. *Choosing Social Safety Net Programmes and Targeting Methods in LDCs*. Washington, D.C.: International Food Policy Research Institute.

Coady, David P., Margaret Grosh, and John Hoddinott. 2004. "Targeting Outcomes Redux," *The World Bank Research Observer*, 19 (1): 61–85.

Deaton, Angus. 1997. *The Analysis of Household Surveys*, Chapter 5. Baltimore: World Bank and Johns Hopkins University Press.

Dutta, Bhaskar and Bharat Ramaswami. 2001. "Targeting and Efficiency in the Public Distribution System: Case of Andhra Pradesh and Maharashtra," May 5–11, *Economic and Political Weekly*, XXXVI (18): 1524–32.

Gelbach, Jonah and Lant Pritchett. 2000. "Indicator Targeting in a Political Economy: Leakier Can be Better," *Journal of Policy Reform*, 4 (2): 113–45.

Government of India. 2002. *Report of the High Level Committee on Long-Term Grain Policy*, Department of Food & Public Distribution, Ministry of Consumer Affairs, Food and Public Distribution.

Howes, Stephen and Shikha Jha. 1992. "Urban Bias in Indian Public Distribution System," *Economic and Political Weekly*, 27 (19; May 9): 1022–30.

Jalan, Jyotsna and Rinku Murgai. 2006. "An Effective 'Targeting Shortcut'? An Assessment of the 2002 Below-Poverty Line Census Method," Centre for Development Economics. Available at http://www.cdedse.org/conf2007/rmurgai.pdf

Javier, Emil (chairperson of the Task Force on Rice and Staples Sector, Congressional Oversight Committee on Agriculture and Fisheries Modernization). 2011. Agriculture and Fisheries 2025: Rice and Other Staples Sector, Mid-year Conference, Philippine Social Science Council, Quezon City, held on October 07, 2011. Available at http://www.cocafm.gov.ph/downloads/

Jensen, Robert and Nolan Miller. 2011. "Do Consumer Price Subsidies Really Improve Nutrition?" *Review of Economics and Statistics*, 93 (4): 1205–23.

Jha, Shikha and P.V. Srinivasan. 2004. "Achieving Food Security in a Cost Effective Way: Implications of Domestic Deregulation and Reform under Liberalized Trade," MTID Discussion Paper No. 67, International Food Policy Research Institute, Washington, D.C.

Kanbur, Ravi. 2009. "Macro Crises and Targeting Transfers to the Poor," mimeo, Cornell University.

Khera, Reetika. 2011. "Trends in Diversion of PDS Grain," Working Paper No. 198, Centre for Development Economics, Delhi School of Economics.

Kotwal, Ashok, Milind Murugkar, and Bharat Ramaswami. 2011. "PDS Forever?" *Economic and Political Weekly*, 46 (21): 72–76.

Lanjouw, Peter and Martin Ravallion. 1999. "Benefit Incidence, Public Spending Reforms and the Timing of Program Capture," *The World Bank Economic Review*, 13 (2): 257–73.

Mehta, Aashish and Shikha Jha. 2009. "Governance and Hunger: A Case Study from the Philippines," Center of Global Studies Working Paper No. 07, University of California-Santa Barbara, California.

———. 2010. "Missing Rice in the Philippines: Measurement and Meaning," Orfalea Center for Global & International Studies Working Paper, University of California, Santa Barbara, August, 2010. Available at http://escholarship.org/uc/item/0w42h334#page-1

NFA. 2006. "2006 Accomplishment Report." Manila, Philippines: National Food Authority.

Olken, B.A. 2006. "Corruption and the Costs of Redistribution: Micro Evidence from Indonesia," *Journal of Public Economics*, 90: 853–70.

Parikh, Kirit S. 1994. "Who Gets How Much from PDS: How Effectively Does It Reach the Poor?" *Sarvekshana*, 17 (3; January–March).

Planning Commission of India. 2005. *Performance Evaluation of Targeted Public Distribution System*, Programme Evaluation Organization, Report No. 189, Planning Commission, Government of India.

Ravallion, Martin. 2009. "How Relevant Is Targeting to the Success of an Antipoverty Program," *The World Bank Research Observer*, 24 (2): 205–31.

Tata Economic Consultancy Services. 1998. "Study to Ascertain the Extent of Diversion of PDS Commodities," Ministry of Food & Consumer Affairs, Government of India.

World Bank. 2001. "Philippines: Filipino Report card on pro-Poor services." Available at http://info.worldbank.org/etools/docs/library/103100/ReportcardsPhilippines.pdf

PRACHI MISHRA*
International Monetary Fund
DEVESH ROY
International Food Policy Research Institute

Explaining Inflation in India: The Role of Food Prices**

ABSTRACT This paper conducts a forensic examination of inflation in India with a focus on food price inflation, using a disaggregated high-frequency commodity-level dataset spanning last two decades. First, we document stylized facts about the behavior of overall, food and nonfood inflation. Second, we quantify the contribution of specific commodities to food price inflation. We find animal source foods (milk, fish), processed food (sugar, edible oils), fruits and vegetables (e.g., onions), and cereals (rice and wheat) to be the primary drivers. Finally, we conduct case studies of the top contributors. Combining the insights from macro as well as micro analyses, the paper suggests specific policy implications.

Keywords: *Food, Inflation*

JEL Classification: *E31, Q18*

Introduction

While inflationary policies can provide a short-run stimulus to the economy, the literature suggests that, over the medium and long term, high rates of inflation are not conducive to economic growth. In a seminal paper, Fischer (1993) presents cross-country evidence to show that growth is negatively associated with inflation, and the causality runs

* We are especially grateful to Arvind Panagariya for extensive comments and Larry Ball for many discussions. We would also like to thank Andy Berg, Olivier Blanchard, Jordi Gali, Peter Pedroni, Isher Judge Ahluwalia, T.N. Srinivasan, Kanhaiya Singh, and participants at the IPF, and the RBI for helpful comments. Enrico Berkes and Manzoor Gill provided excellent research assistance.

** The views expressed here are those of the authors and do not necessarily represent those of the IMF/IFPRI or IMF/IFPRI policy.

from inflation to growth. The mechanisms through which high inflation can reduce growth are through declines in investment and productivity growth. He also presents few exceptional cases to show that low inflation (<15 percent) is not necessary for high growth, but that high inflation (>50 percent) is not consistent with sustained growth. Overall, he concludes based on the regression evidence and case studies that low inflation is conducive to sustained growth.[1]

Furthermore, it is often claimed that the costs of inflation are likely to be borne most heavily by the poor. The underlying argument is that the poor are typically wage earners, and their incomes change slowly. As argued in Fischer and Modigliani (1978), "it has typically been believed that wages lag behind inflation, and that inflation therefore implies a shift away from wage earners, and towards profits." In addition, the rich have assets that are more likely to be indexed to inflation. According to Easterly and Fischer (2001), "a growing body of literature on balance—but not unanimously—tends to support the view that inflation is a cruel tax." They present cross-country evidence to show that improvements in the poor's share of national income, the percentage decline in poverty and the percentage change in the real minimum wage are all negatively correlated with the rate of inflation. In the case of India, Ravallion and Datt (1996) use data across states and over time to show that higher inflation is associated with higher poverty rates.

Both from the perspective of its effect on growth and its impact on the poor, inflation remains a top concern in India. The current Prime Minister of India noted recently that "inflation poses a serious threat to the growth momentum. Whatever be the cause, the fact remains that inflation is something which needs to be tackled with great urgency." In terms of the recent trend in inflation, India does stand out among the emerging economies. The Consumer Price Index (CPI) inflation rate reached 12 percent in 2010, a much higher level than the group of other emerging economies commonly known as BRICS (Brazil, Russia, India, China, and South Africa) (Figure 1). Based on the officially reported CPI figures, the inflation rate in India in 2010 was close to thrice that in China and South Africa, and roughly twice the levels in Brazil and Russia. Moreover, the trend in the inflation rate in

1. There is a vast literature which suggests that the relationship between inflation and growth tends to be nonlinear. However, there is significant debate about the cutoff point beyond which inflation becomes detrimental to growth. For example, Bruno and Easterly (1995) suggest 40 percent as a danger point, beyond which increases in inflation are very likely to lead to a growth crisis. In contrast, Khan and Senhadji (2000) estimate that the threshold above which inflation significantly slows growth is 1–3 percent for industrial countries and 7–11 percent for developing countries.

FIGURE 1. Inflation in Emerging Economies

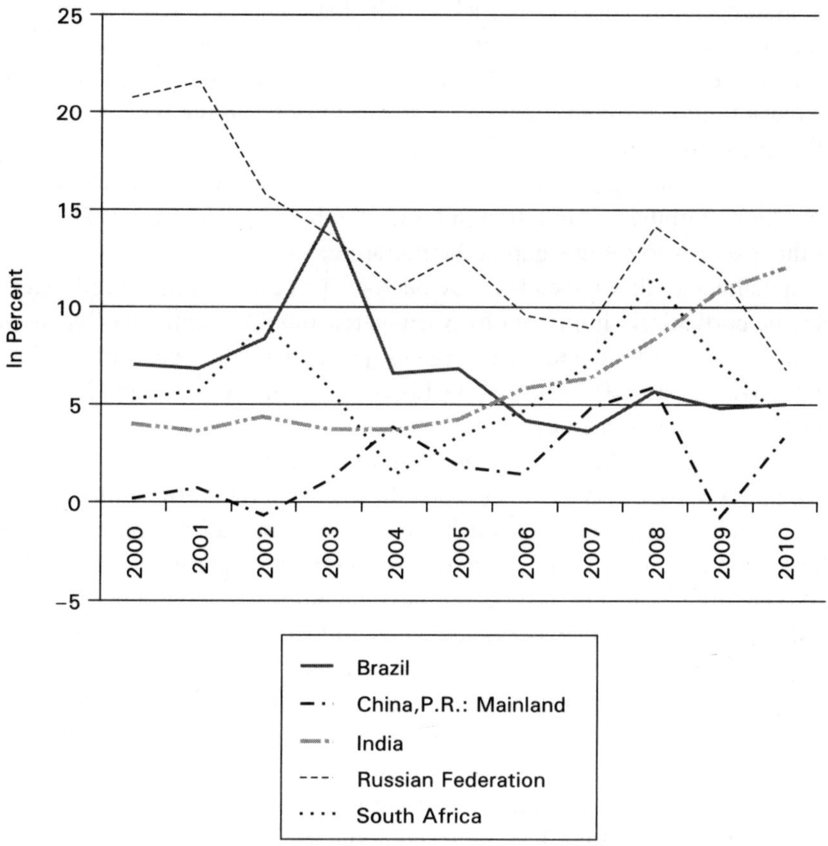

Source: International Financial Statistics, IMF.
Notes: This graph shows the annual inflation rates based on the consumer price indices (CPI).

India is striking. While the other BRICS show more or less a declining trend during the last decade, India has had a steady upward trend.

Given the recent episodes of high inflation, it is important to investigate its primary drivers, e.g., food, nonfood, etc. From a policy perspective, to tackle inflation, it may, however, not be sufficient to identify the sources only at a broad level, particularly in the context of a developing country like India. In India, both macro policies like monetary tightening by the RBI as well as commodity-specific measures by different branches of the government (e.g., trade policies and domestic interventions) have been used to deal with inflation. Therefore, combining the perspectives from both macro as well as micro data may be crucial to get a better diagnosis and to design policies to rein in inflation. In particular, understanding the heterogeneities across commodities could be important to inform macro policy. For example, the assumption in standard macro models that changes in relative price of food and fuel represent supply shocks may not hold for many commodities. Furthermore, given that there has been a persistent upward trend in inflation, taking a long-term view rather than focusing only on the recent inflation episodes seems imperative.

To our knowledge, there is little systematic empirical evidence on the long-term evolution of inflation in India at a disaggregated level. The goal of this paper is to fill this gap in the literature.

Specifically, this paper focuses on food prices. Studying food prices may be particularly important for various reasons. As argued by Cecchetti (2007), ignoring food and energy prices, particularly in recent times when they have consistently been rising faster than other prices, could lead to biased estimates of medium-term inflation.

In the case of India with a large share of population that is poor (42 percent in 2005, based on Chen and Ravallion, 2010), food price inflation can be particularly important. This is because poor spend a large proportion of their income on food (over 50 percent based on 2004–05 round of the NSS), are typically net buyers of food and have incomes that tend to be fixed.[2]

The rest of the paper is organized as follows. Section "Measurement of Inflation in India" discusses the measurement of inflation in India, and section "Trends in Overall Inflation" discusses the trends in overall inflation. Section "How Important Is Food in Overall Inflation?" establishes the importance of food, section "Which Commodities Drive Food Price Inflation?" identifies the sources of food price inflation at the commodity level, and section "What Factors Explain the Rise in Food Price Inflation? Case Studies of Specific Commodities" presents case studies of specific commodities. Section "Conclusions and Policy Implications" concludes and presents some policy implications.

Measurement of Inflation in India

Price Indices and Inflation

The official price indices in India are calculated using the Laspeyres formula. Formally, the index is calculated as,

2. Panagariya (2005) argues more generally that an increase in food prices due to removal of OECD subsidies might actually hurt the poor in developing countries, many of whom tend to be net buyers. Specifically in the Indian case, de Janvry and Sadoulet (2008) estimate a large share of the rural population to be net food buyers.

$$I_t = \frac{\sum_{j=1}^{n} I_{jt} W_{j0}}{\sum_{j=1}^{n} W_{j0}} \tag{1}$$

where I_{jt} is the price index of the jth commodity/item at time t, W_{jo} is the weight assigned to jth item in the overall price index. The weight of jth item is given by $\frac{W_{j0}}{\sum_{j=1}^{n} W_{j0}}$, where W_{j0} is the expenditure on item calculated using base year prices and quantities.[3]

The Laspeyres price index can in principle be calculated using producers', consumers', or wholesale prices. The index calculated at each of the three sets of prices serves different purposes. The Producer Price Index (PPI) measures the average selling prices received by domestic producers of goods and services. This contrasts with other measures, such as the CPI, that measure average prices from the consumer's perspective. Sellers' and consumers' prices may differ, e.g., due to taxes and subsidies, and distribution costs. Finally, the Wholesale Price Index (WPI) ideally measures the average prices in the wholesale market, i.e., where goods are sold in bulk. These price indices are used to measure the average change over time in the selling prices received by producers (PPI-inflation) or prices paid by consumers (CPI-inflation), or the average price change in the wholesale market (WPI-inflation).

The WPI is the most commonly used price index in India. The term "wholesale" in the index is, however, misleading in that the index does not necessarily measure prices in the wholesale market. In practice, the WPI in India measures prices at different stages of the value chain. As discussed in Srinivasan (2008), according to the National Statistical Commission (NSC, 2001), "in many cases, these prices correspond to farm-gate, factory-gate or mine-head prices; and in many other cases, they refer to prices at the level of primary markets, secondary markets or other wholesale or retail markets."

The weights used in the WPI are revised every decade. The latest series available is based on 2004–05 as the base year, and includes 676 commodities. The weights used in the 2004–05 = 100 index are shown in Figure 2. Food has a total weight of 24.3 percent—14.3 percent allocated to primary

3. I_{jt} is arrived at as the simple arithmetic average of the price relative $(P_1jt/P1j0 * 100)$ of all the varieties of commodity j. Hence equation (1) can also be expressed as $I_t = \frac{\sum_{j=1}^{n} P_{jt} q_{j0}}{\sum_{j=1}^{n} P_{jt} q_{j0}}$, where q_{j0} is the base period quantity of commodity j.

FIGURE 2. Weights. WPI, 2004–05 = 100

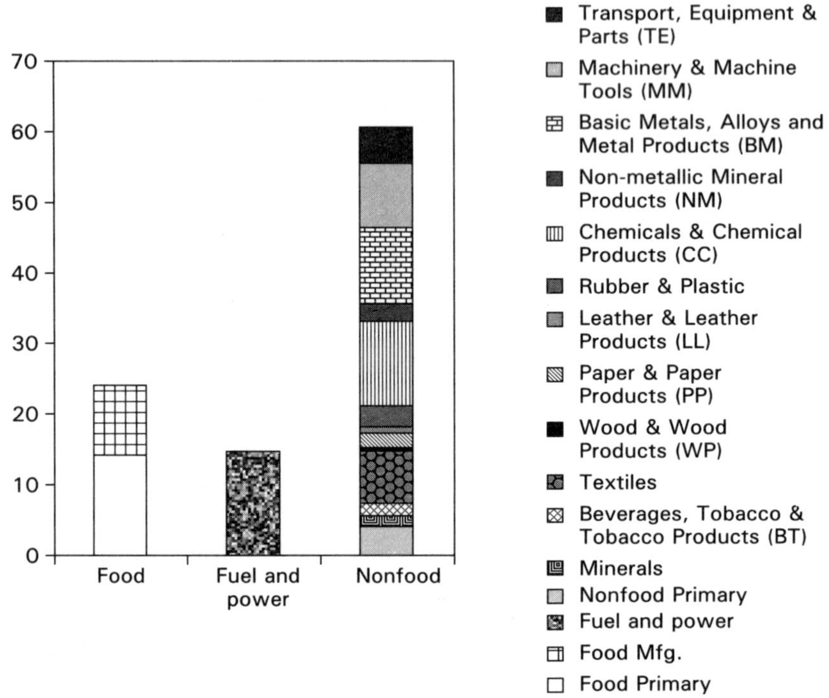

Source: Ministry of Commerce and Industry, Government of India.

food articles, and the rest to manufactured ones. Fuel has a weight close to 15 percent. Among the nonfood, nonfuel category, the three largest weights are for chemicals (12 percent), metals (10.7 percent), and textiles (7.3 percent). The WPI does not include any services.

In addition to the WPI, four consumer price indices are also officially published—these correspond to different segments of the population, namely, industrial workers (CPI-IW, base year 2001), agricultural laborers (CPI-AL, base year 1986–87), rural laborers (CPI-RL, base year 1986–87) and urban nonmanual employees (CPI-UNME, base year 2001). A nationwide measure of CPI that combines rural and urban areas has become available only recently starting January, 2011 (base year 2010). The weights used for new CPI have been derived from the 61st (2004–05) round of the NSS.[4] The

4. The CPI weights for CPI-AL and CPI-RL are based on consumption expenditures from the National Sample Survey Organization (NSSO) surveys; whereas those for CPI-IW and CPI-UNME are based on ad hoc family expenditures in selected urban centers only (Srinivasan, 2008).

officially published CPI-IW include only six sub-indices—clothing, food, fuel and light, housing, tobacco and intoxicants, and miscellaneous. CPI-AL and CPI-RL are published with only five subindices, and exclude housing. The commodity coverage for the latest CPI measures has been broadened, and includes 23 items. In addition to housing, six subcategories of services have been included. Importantly, CPI gives a much larger weight to food compared to WPI (Figure 3, the weight on food in CPI ranges from 46 to 69 percent depending on the segment of workers the index refers to), hence is more likely to capture the recent surges in food prices.[5]

The CPI is arguably a better measure than WPI to study changes in prices of final goods demanded by consumers. The analysis in this paper, however, is based primarily on WPI instead of CPI due to three main reasons. First, historical data on the nationwide measure of CPI does not exist. Second, CPI is not available at a disaggregated commodity level (even the most recently released nationwide index is much more aggregated compared to the WPI). Finally, CPI is as yet not commonly used for policy purposes. Nonetheless, we also analyze CPI, wherever possible for comparison.

Several papers have pointed out the deficiencies of the price indices used in India for measuring inflation (see Srinivasan, 2008). The recommendations to improve the indices include converting WPI into a PPI and also to expand the scope of WPI to include services, and of the CPI (2010 = 100) to expand the coverage of services. The CPI, even in its latest incarnation (2010 = 100) has a limited coverage of both goods and services (total of 23 items), hence is not amenable to any serious sectoral or disaggregated analysis. Therefore, if CPI were to become the primary index for policy purposes in the longer term (as is the norm in most countries), expanding its coverage seems imperative.

We use WPI data at the monthly frequency, covering the period July 1988–May 2011. The 2004–05 = 100 index with 676 items is only available for the period starting April 2004. In order to create one comparable series for the WPI over the entire period from 1988 to 2011, we project the 2004–05 series backwards using the growth rates in the price indices based on the 1993–94 (including 435 commodities), and 1981–82 series (covering 447 commodities). Although we confirmed that using either series does not significantly change the trends in overall inflation, yet appropriate caution should be exercised in comparing *aggregate* inflation trends based on the changing basket of commodities over time.

5. The weight on fuel is lower in the CPI than in WPI (6.4 percent in CPI-IW).

FIGURE 3. Weights in the Food Basket: WPI (2004–05 = 100)

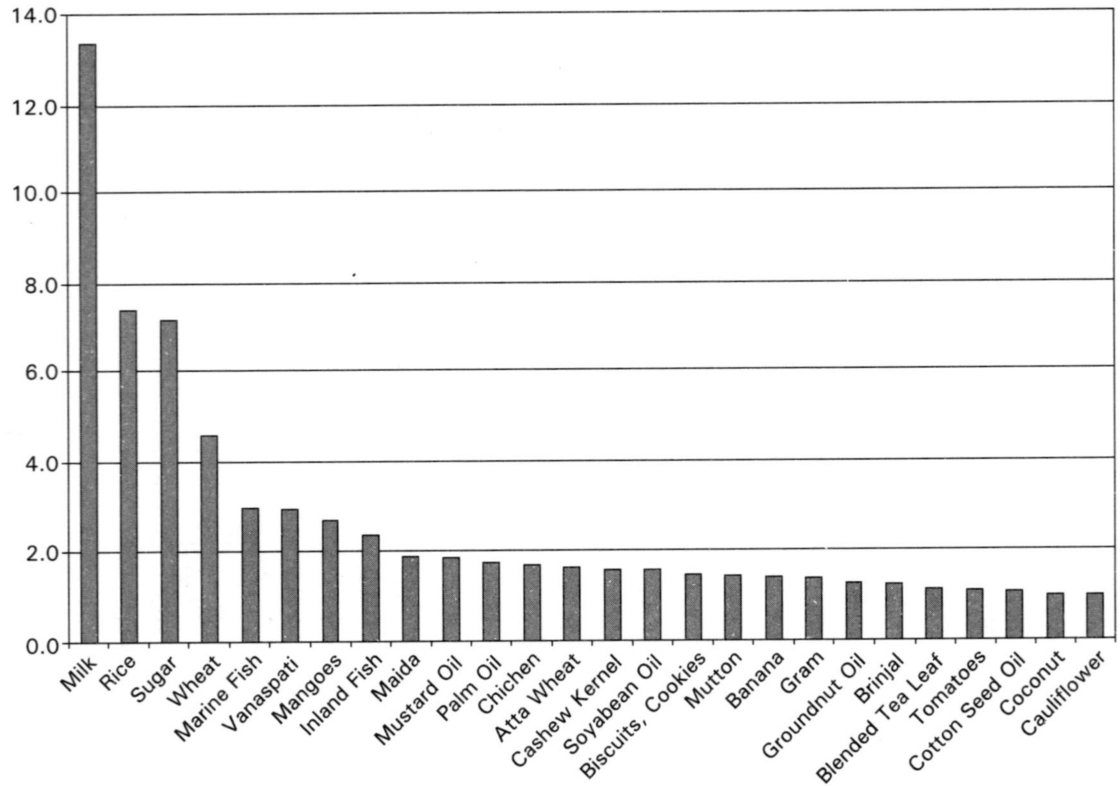

Source: Ministry of Commerce and Industry, Government of India.

Food Prices and Overall Inflation

As any price index is composed of many goods and prices, we need to distinguish between situations where all prices change in the same proportion (absolute-price changes or "pure inflation") and changes in prices of some goods relative to others (relative-price changes).[6] According to the classical theory, relative price changes are determined by real factors, whereas changes in money supply drive absolute prices. Friedman argued that changes in prices of a subset of commodities, e.g., food or oil, should not feed into aggregate inflation. A rise in the relative price of food (for a given level of money supply) should make consumers spend more on food, and less on other items, keeping the aggregate price level and the inflation rate unchanged.

There has been a vast literature thereafter which has studied the links between relative price changes and aggregate inflation. For example, Fischer (1982) introduces the hypothesis of downward inflexibility of prices to explain why relative price changes could feed into aggregate inflation. Subsequently, Ball and Mankiw (1995) show that when price adjustment is costly (through menu costs) and firms adjust to large shocks but not to small shocks, aggregate inflation depends on the distribution of relative price changes: inflation rises when distribution is skewed to the right. Therefore unlike the implicit assumption of perfect flexibility of all nominal prices in Friedman's analysis, in a real world with rigidities, a link between relative price changes and aggregate inflation is clearly possible. Such a link motivates analysis of price changes in a subset of commodities.

The price of any subset of commodities can be expressed using a formula similar to Equation (1). Specifically, this paper focuses on food prices for reasons discussed earlier.

The food price index can be specified as

$$I_t^F = \frac{\sum_j (W_{j0}^F * I_{jt}^F)}{\sum_j W_{j0}^F} \tag{2}$$

where I_{jt}^F is the price index of the jth food item at time t, W_{j0}^F is the weight assigned to jth food item in the food price index (not the overall price index).

6. The concept of pure inflation goes back to David Hume (1752): "imagine all prices increase in the same proportion, but no relative price changes." See also Blanchard (2009) for definition of pure inflation and Reis and Watson (2010) for an analysis in the case of United States of both absolute and relative price changes.

The weight of *j*th food item is given by $\frac{W_{j0}^F}{\sum_{j=1}^{n} W_{j0}^F}$, where W_{j0}^F is the expenditure on *j*th food item calculated using base year prices and quantities.

In order to identify the specific contributors to food price inflation, we also compile data on prices at the disaggregated commodity level, which are available only from April 1994. There are in total 112 primary and manufactured food items in 2004–05 = 100 index. The top 25 items in terms of weights are shown in Figure 4. Milk has the highest weight in the food basket (13.3 percent) followed by rice (7.4 percent), sugar (7.4 percent), wheat (4.6 percent), and marine fish (3 percent).[7]

FIGURE 4. Weights: CPI, 2001 = 100

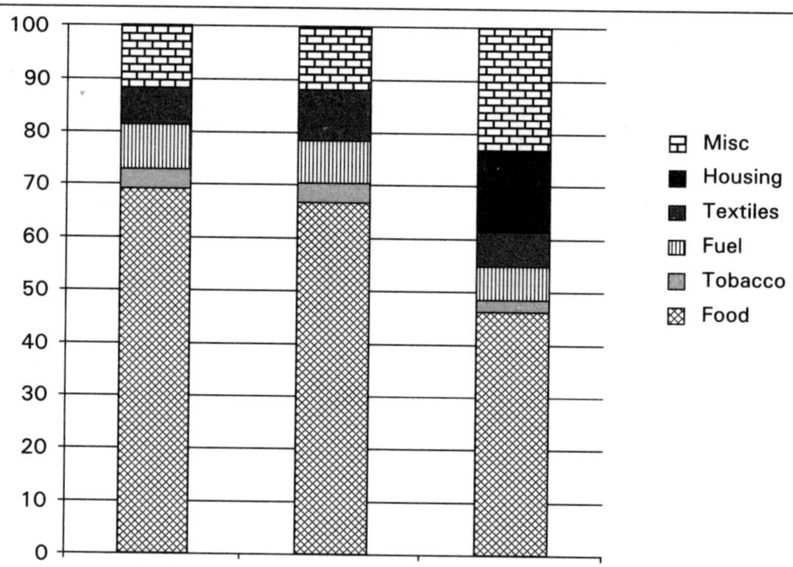

Source: Ministry of Commerce and Industry, Government of India.

All price data is publicly available on the website of the Ministry of Commerce and Industry, Government of India (GOI). We rebase all the data to the year 2005 = 100. Inflation is calculated on a y-o-y basis. For the time series analysis, however, we use quarterly data seasonally adjusted

7. The weights at the disaggregated commodity-level are not published online for the 1993–94 and 1981–82 indices; the weights for aggregate categories are available from a report from the Ministry of Commerce and Industry.

using the X12 method.[8,9] Finally, although food and fuel prices may show similar dynamics, the focus of this paper is on food prices, and hence we will exclude fuel; the term "nonfood" will denote the items excluding food as well as fuel.

Trends in Overall Inflation

The y-o-y rates of change of different price indices (the four CPIs, and the WPI) are shown in Figure 5. The four CPI inflation rates are highly correlated with each other (bivariate correlations of 0.86 or higher). However, these are weakly correlated with the WPI-inflation rate. The correlations of both CPI-AL and CPI-RL with WPI equal 0.25, whereas CPI-IW and CPI-UNME have a comparatively high correlation of 0.6.

Since the different CPIs are highly correlated with each other, for a comparative analysis of CPI with WPI, we focus only on CPI-IW (here-after CPI). The historical evolution of average y-o-y inflation (both CPI and WPI) is shown in Figure 6. There are three peaks in CPI inflation during the last two decades. First, the inflation rate reached 14.1 percent in 1991, when India had the balance of payments crisis; the next peak of 13.2 percent occurred in 1998. Finally, the third happened in 2010 with the inflation rate mounting to 12.1 percent, the highest during the last decade. Importantly, as will be shown in the next section, a sharp rise in food prices contributed to all the three episodes. Notice that while the CPI and WPI inflation rates were close to each other during the 1991 peak; they diverged significantly around the latter two peaks. The higher weight of food in the CPI basket and that of fuel in the WPI explains these patterns. In 1991, while the increase in the food inflation rate contributed the most to the CPI inflation, fuel inflation also increased significantly. In contrast, during 1998 and 2009, while there was a sharp collapse in the fuel inflation rate, food prices inflation continued to rise steadily.

8. Although monthly series on y-o-y inflation are reasonable for presentation purposes, they can exaggerate the degree of persistence. Therefore, for econometric analysis, we use quarterly data as is standard in many studies (e.g., see the series of studies conducted by the ECB at http://www.ecb.int/events/conferences/html/inflationpersistence.en.html).

9. We recognize that the X12 methodology may not be appropriate to adjust for seasonality in the Indian context (e.g., due to the fact that certain festivals need not fall in the same quarter every year). However, the government does not publish seasonally adjusted price data. Therefore, we repeat the analyses in the paper also with data that is not seasonally adjusted, and the results are qualitatively similar.

FIGURE 5. Various Prices Indices in India: Rates of Change

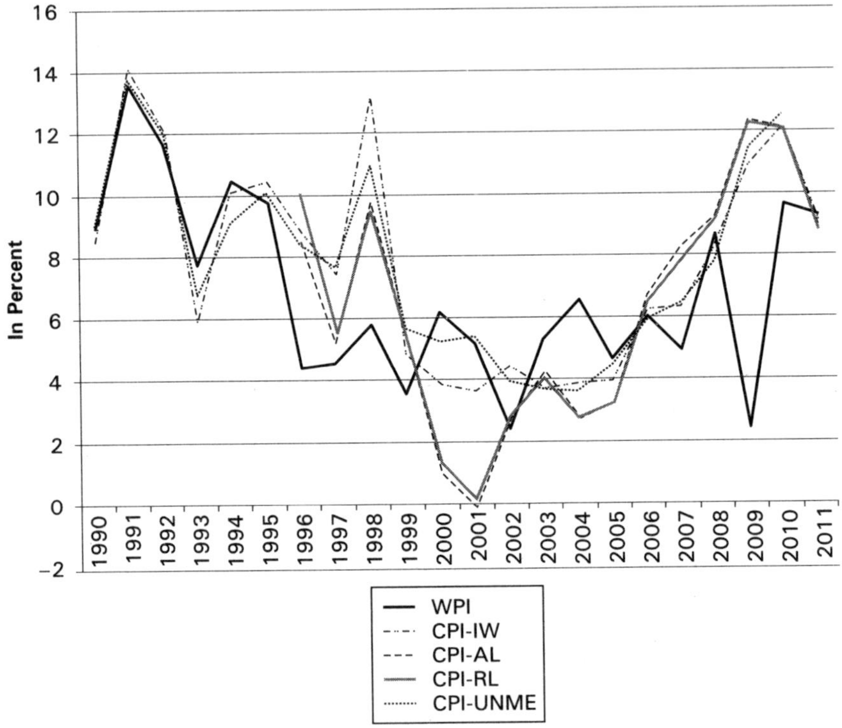

Source: Ministry of Commerce and Industry, Government of India.

Notes: This figure shows the average y-o-y inflation rates. The WPI figure for 2011 is based on Jan–May; for CPI, from Jan–April. CPI-IW, CPI-AL, CPI-RL, and CPI-UNME are the CPI for industrial, agriculture, rural, and urban nonmanual employees, respectively.

In order to assess the underlying trends in inflation, we use a Hodrick-Prescott (HP) filter (assuming a smoothing parameter of 100—traditional value for annual frequency data). The trend inflation rate based on both CPI and WPI show a U-shaped pattern (Figure 6). While there was a downward trend in inflation in the 1990s; this has reversed since the early 2000s, whereby the inflation rate based on either index shows a clear upward trend. The Bai and Perron (1998) test for structural break identifies a break in the trend in 2000, and an inflection point in 2002.

Subsequently, we divide the data into five subperiods, 1990–94, 1995–99, 2000–04, 2005–09, 2010–11, and looked at the average inflation rates during these periods. Strikingly, there has not been a single subperiod with average WPI-inflation rate less than 5 percent. In fact, the average WPI inflation rate was higher than 9 percent in two subperiods: 1990–94 and 2010–11.

FIGURE 6. Inflation Rate: CPI and WPI

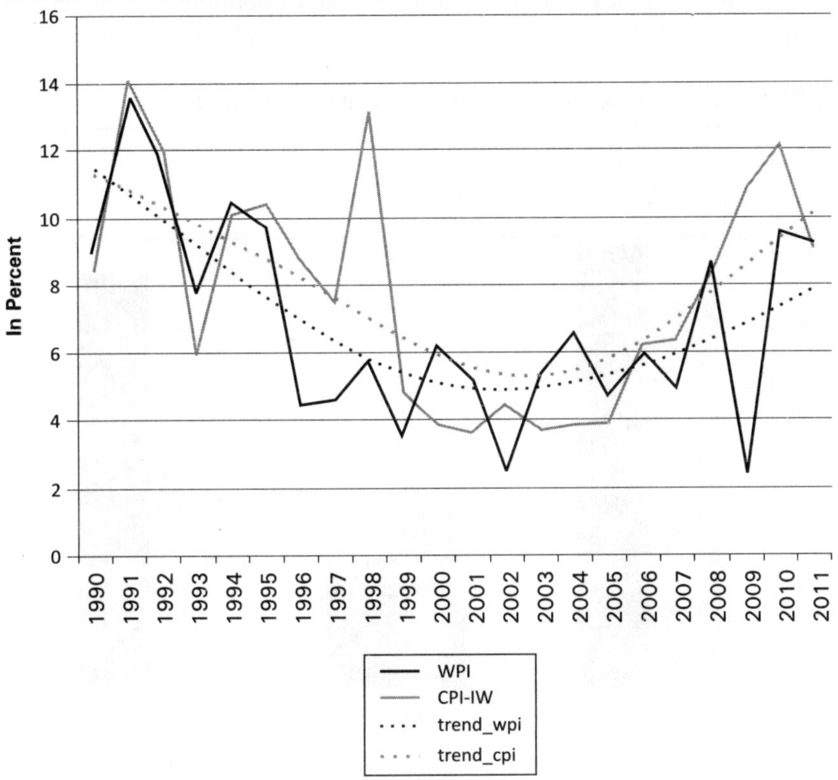

Sources: Ministry of Commerce and Industry, Government of India, and authors' calculations.

Notes: This figure shows the average y-o-y inflation rates.The WPI figure for 2011 is based on Jan–May; for CPI, from Jan–April. The trends are estimated based on a HP filter, with a smoothing parameter of 100.

In case of CPI, there has been only one subperiod, 2000–04, when the inflation rate was less than 5 percent. Food price inflation (discussed in detail below) tracks CPI-inflation and was lower than 5 percent only in 2000–04. In all the other subperiods, food price inflation was more than 7 percent.

Further, we conduct tests for multiple structural breaks based on Bai and Perron (1998).[10] The tests reveal two breaks in the mean of overall inflation in 1999 and 2006. Accordingly, we create three subperiods: 1990–98, 1999–2005, and 2006–11. The average WPI inflation rate is highest in the

10. We conduct the tests using monthly data. More precisely, the breaks occur in April 1999 and April 2006. The test is implemented using R software package. For implementation details, see Zeileis et al., 2003.

first segment and the second segment is a period of relatively low inflation before it picks up again in the third subperiod. Importantly, the patterns of food price inflation are congruent with that of overall inflation (Figure 7).

FIGURE 7. Average Inflation Rate by Sub-periods (in percent)

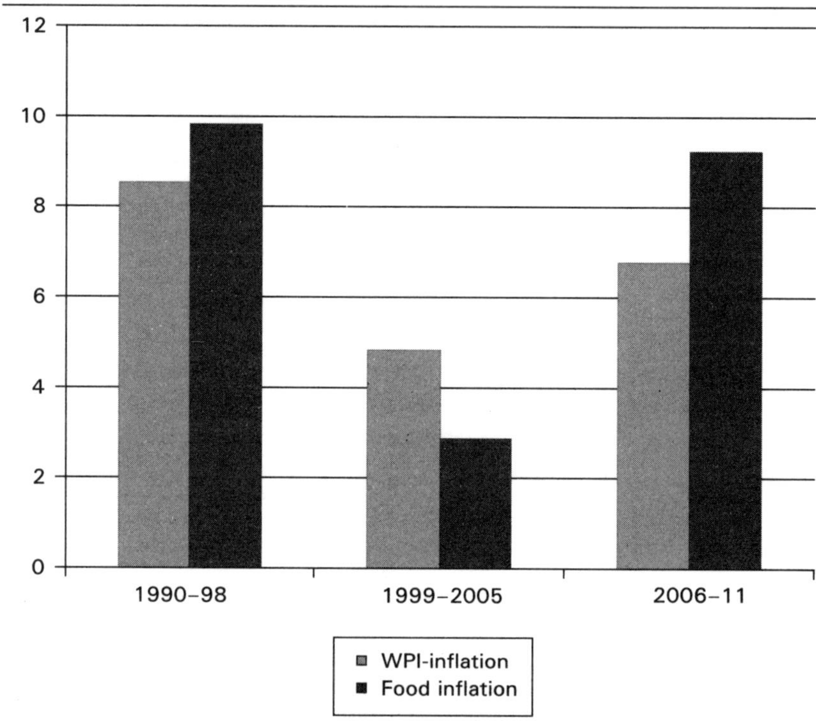

An important point to note is that low inflation (defined as 5 percent or less) has been a rare occurrence in the Indian economy over the last two decades.[11] Based on WPI, there have been only seven years when the inflation rate was less than 5 percent. Strikingly, not even once has less than 5 percent inflation rate persisted for more than two years.[12] Also, in the case of CPI, there have been only seven years with low inflation; the difference from WPI is that these seven years have been in continuum starting from 1999 (with food prices being less than 5 percent in all these years).

11. The RBI's implicit objective is lower than 5 percent. Its objective is to "condition and contain perceptions of inflation in the range of 4.0–4.5 percent, in line with the medium-term objective of 3.0 percent inflation …" (Pattanaik and Nadhanael, 2011).

12. Only in 1996 and 1997, the WPI inflation rate was consistently below 5 percent.

Although inflation in 2011 (particularly the CPI inflation) seems to have declined from its 2010 peak, mainly due to easing of pressure on food prices; inflation continues to remain a source of concern due to two reasons: (*a*) the rate is still high and close to double digits, and (*b*) there are no signs of a downward *trend*.

Any discussion on inflation may be incomplete without looking at evolution of monetary policy. Figure 8 shows the movement of the policy rate (or the repo rate, the rate at which RBI lends money to commercial banks, which is the primary instrument for conducting monetary policy in India).[13] Although there have been some fluctuations, the long-term trend

FIGURE 8. Inflation and Policy Rate

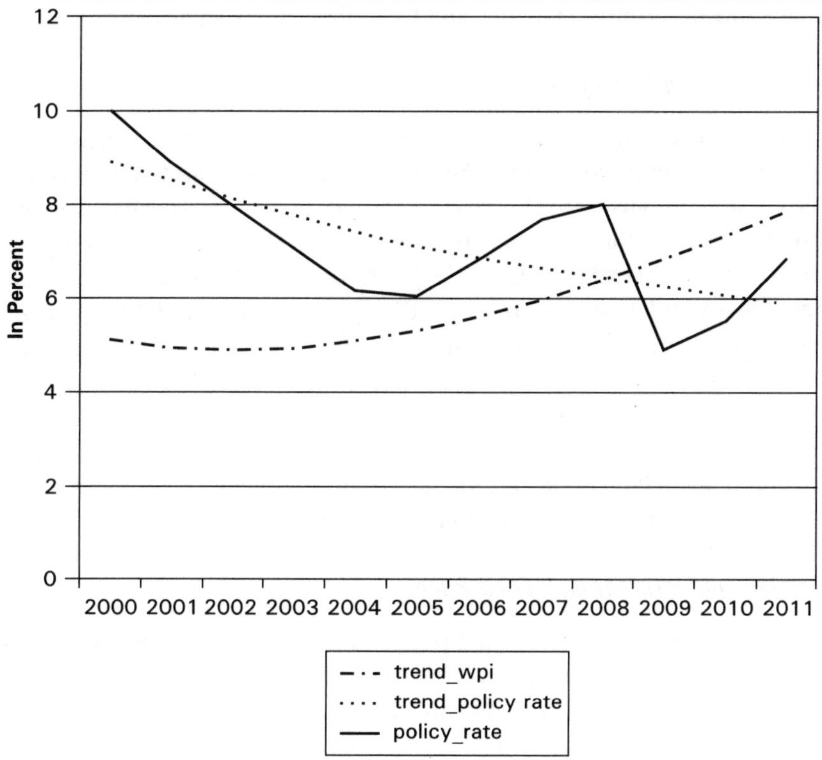

Sources: Ministry of Commerce and Industry, Government of India, and Reserve Bank of India.

Notes: This figure shows the trend in the average y-o-y inflation rates based on WPI. The policy rate is the repo rate, the rate at which the RBI lends money to commercial banks. WPI figure for 2011 is based on Jan–May; for repo rate, from January–June. The trends are estimated based on a HP filter, with a smoothing parameter of 100.

13. Data on the repo rate is available from the last quarter of 2000.

in the policy rate has clearly been downward, which suggests that monetary policy has largely been loose over the last decade. In fact, since 2009 real interest rates have been negative. Although one cannot claim that monetary policy "caused" the upward trend in inflation, there certainly appears to be a strong positive association between the two.

How Important Is Food in Overall Inflation?

In this section, we delve into the behavior of food price inflation and its relationship with nonfood and aggregate inflation. The goal of this section is to systematically document stylized facts about the importance of food, using some descriptive statistics and basic time series analysis.

Broad Trends in Food Inflation

We start by discussing some broad trends in food price inflation. We calculate food price inflation combining the indices for primary and manufactured food commodities using the Laspeyre's formula and their respective weights. Looking at the historical trends, there have been three peaks of food price inflation in the last two decades—1991, 1998, and 2010—where the average annual y-o-y inflation rates reached 17.9 percent, 11.1, and 14.2 percent, respectively (Figure 9). In fact, if we look at the peaks of the monthly y-o-y inflation reached during a given year, the three peaks were higher—20.8 percent in 1991, 18.1 percent in 1998, and 20.2 percent in 2009–10 (Figure 10A). While we will look closely at the contributors of food price inflation in the latter two peaks, the lack of disaggregated commodity-level data on prices prior to 1994 precludes a deeper analysis for 1991.

Domestic and International Food Prices

In this subsection, we make a first pass at the question: how integrated are domestic food prices with their international counterparts. This question has become particularly relevant since the global food price crisis in 2008.

We find only a moderate correlation between domestic and international food price inflation (close to 0.5, Figure 11A).[14] At the commodity level, there is significant variation in the degree of co-movement between domestic and international prices (Table 1). In general, the relationship is weaker for staples like rice and wheat compared to other items like edible oils and

14. This is consistent with research done at the RBI (e.g., see Rajmal and Misra, 2009).

FIGURE 9. Inflation Rate: Food and Nonfood

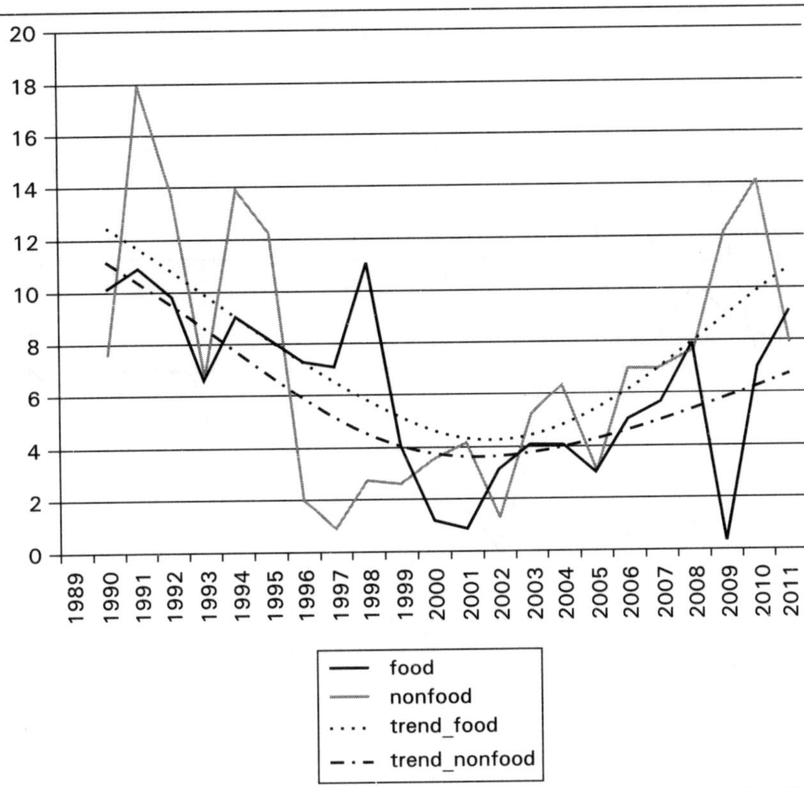

Sources: Ministry of Commerce and Industry, Government of India, and authors' calculations.

Notes: This figure shows the average y-o-y inflation rates based on WPI. The trends are estimated based on a HP filter, with a smoothing parameter of 100.

sugar. This may reflect, in part, the government's reluctance to allow any significant pass-through from international to domestic prices in the case of staples (as discussed in detail in section "What Factors Explain the Rise in Food Price Inflation? Case Studies of Specific Commodities"). Broadly, the degree of co-movement depends on the actual or potential tradability of the commodity.[15] Highly tradable products like edible oils therefore exhibit a high degree of co-movement.

Another interesting and consistent pattern that emerges across commodities is that the degree of co-movement between domestic and international

15. For example, if an export ban is in place on a potentially exportable commodity, then a rise in international prices can put an upward pressure on domestic prices through different channels, e.g., illegal trade, or political pressure to raise support prices.

FIGURE 10A. Peaks. Food and Nonfood Inflation

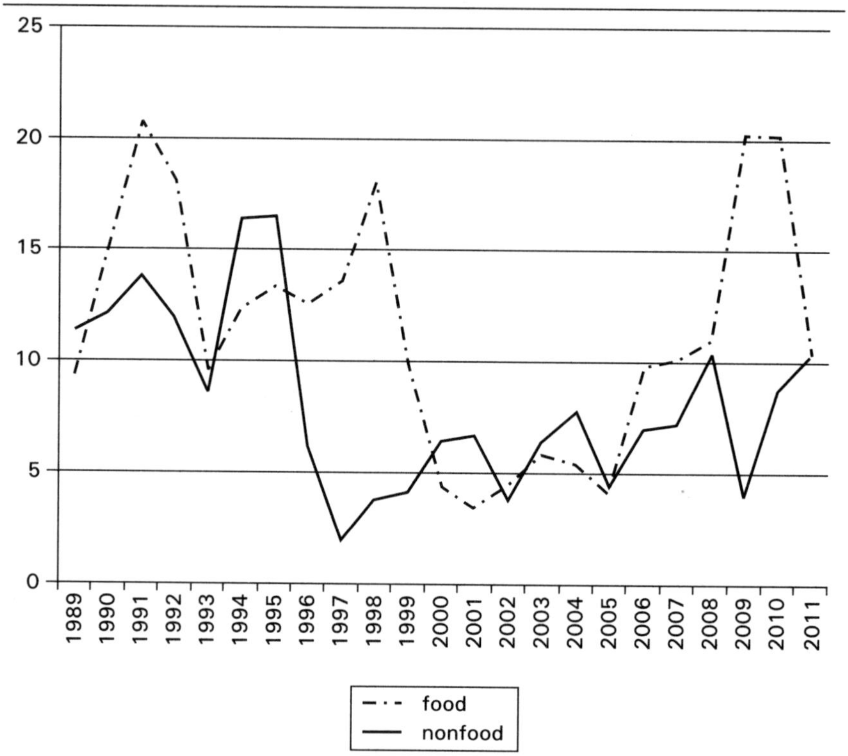

Sources: Ministry of Commerce and Industry, Government of India, and authors' calculations.

Notes: This figure shows the peaks in any month in a year for the y-o-y inflation rates based on WPI. The WPI figure for 2011 is based on Jan–May; for CPI, from Jan–April.

TABLE 1. Correlation between Domestic and World Food Price Inflation

Overall	**0.47**
Rice	0.14
Wheat	0.30
Maize	0.15
Sugar	0.62
Edible oils	0.88
Chicken	0.32
Fish	0.55

Notes: The world prices are taken from the Commodities Database at the IMF, and covers 13 food items. The domestic food price inflation is based on a weighted average of WPI for the same 13 items.

prices is stronger when international prices are low than when they are high (e.g., in the case of rice and wheat shown in Figures 11B and 11C). This may suggest that the government is more unwilling to allow the pass-through, when prices are higher.[16]

Comparing Food and Nonfood Inflation

How does the magnitude of food price inflation compare with that of nonfood (core)? This question is important to evaluate whether policymakers can afford to ignore food in deciding which measure of inflation to target.

The instances of food price inflation exceeding nonfood are fairly common during the last two decades. Since 1990, there were 12 such episodes, with the difference between food and nonfood inflation reaching a peak of 11.8 percentage points in 2009 (Figure 9). The year 2010 also recorded a positive differential between food and nonfood inflation of 7.2 percentage points. Figure 9 also shows the HP-filtered trends in food and nonfood inflation. It is striking how the trends are similar with both exhibiting a U-shaped relationship and also moving together. Although both inflation rates show a clear upward trend since the early 2000s, the trend growth in food prices has always been higher than nonfood prices.

The differential between food and nonfood is even starker when one looks at the monthly figures on y-o-y inflation. For example, the difference between the peaks of food and nonfood inflation reached in any month during a year was higher than 16 percentage points in 2009, and close to 11.5 percentage points in 2010 (Figure 10a). Furthermore, there have been only a couple of years during the last two decades with food not exceeding nonfood in at least one month in a year (Figure 10B). 1991, 1997, 1998, and 2009 were the four years in the last two decades when food inflation exceeded nonfood in every single month.

Time Series Properties of Food and Nonfood Inflation

In this subsection, we dig further into the importance of food and nonfood inflation. We start by estimating a simple autoregressive process using quarterly data, where the food and nonfood inflation rates, measured as quarter-on-quarter changes, are regressed on their lags. The optimal lag length is selected by using the Akaike Information Criterion (AIC). The results are

16. One relevant question is whether RBI pursues a policy of monetary accommodation in response to rising international food prices. We did not find any documentation for this. We also looked at the correlation between repo rates and international food prices—the correlation coefficient is positive but small.

FIGURE 10B. Number of Months with Food Inflation Higher than Nonfood

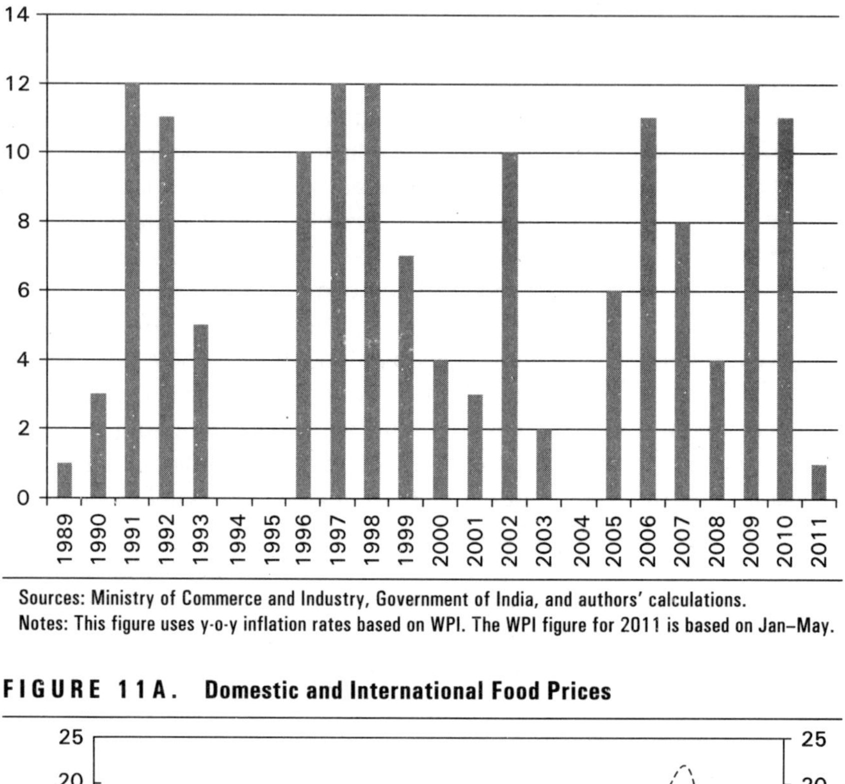

Sources: Ministry of Commerce and Industry, Government of India, and authors' calculations.
Notes: This figure uses y-o-y inflation rates based on WPI. The WPI figure for 2011 is based on Jan–May.

FIGURE 11A. Domestic and International Food Prices

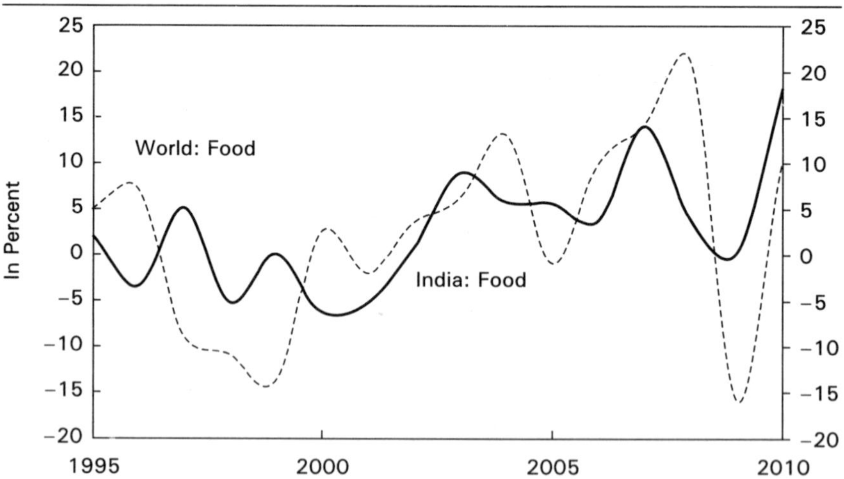

Sources: IMF Commodities Database, Ministry of Commerce and Industry, Government of India and authors'calculations.
Notes: This figure shows the evolution of world and domestic food prices. The world prices are for 13 food items; domestic prices are based on the WPI, and are a weighted average of the 13 items for which the world prices exist.

FIGURE 11B. Rice: Domestic and International Prices

Price Indices (2005 = 100)

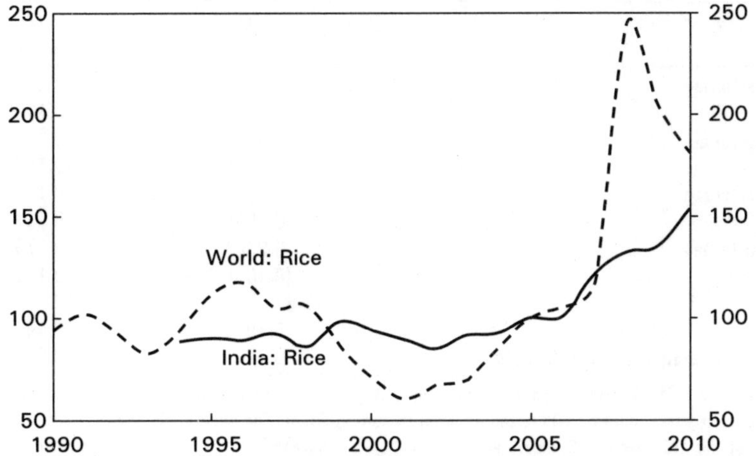

FIGURE 11C. Wheat: Domestic and International Prices

Price Indices (2005 = 100)

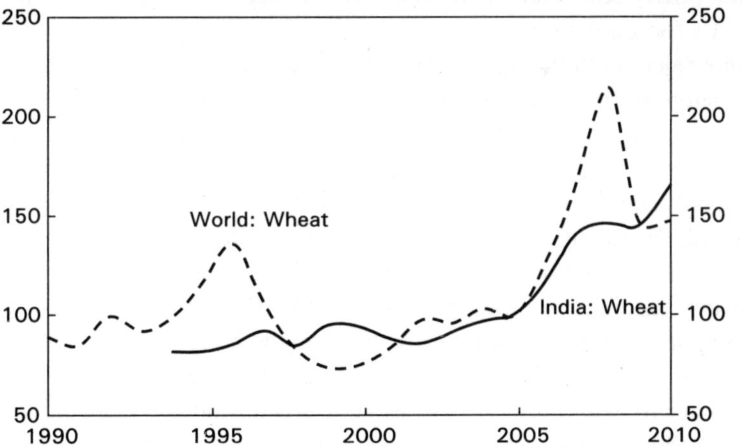

shown in Table 2. The sum of the autoregressive coefficients which is a commonly used measure of persistence, is quite high for food price inflation (about two-thirds of the persistence in nonfood inflation).[17]

17. The long run response to a shock to inflation is given by $\frac{1}{1-\sum \cdot \alpha_j}$, where α_j's are the coefficients on the lags. Hence, larger the sum of autoregressive coefficients, bigger is the response.

TABLE 2. Persistence of Food/Nonfood Inflation: Sum of Autoregressive Coefficients

	[1] Food inflation	[2] Nonfood inflation
1-quarter lag	0.326***	0.607***
	[0.108]	[0.113]
2-quarter lag	0.076	−0.007
	[0.110]	[0.134]
3-quarter lag	0.216*	0.157
	[0.109]	[0.134]
4-quarter lag	−0.23C**	−0.173
	[0.105]	[0.115]
N	87	87
r2	0.19	0.40
Sum of autoregressive coefficients	0.39	0.58

Note: The table reports the coefficients from OLS regressions of food and nonfood inflation calculated based on quarterly data on WPI-based inflation (deseasonalized). The optimal lag length is selected on the basis of the AIC criterion. Standard errors are denoted in parentheses. ***, **, and * denote statistical significance at 1, 5, and 10 percent, respectively.

Next, we test for unit roots in both the series using an Augmented Dickey-Fuller (ADF) test (Table 3A). The first two tests are for the null hypotheses that the food inflation and nonfood inflation series contain a unit root. We cannot reject both the hypotheses at the 1 percent level.[18] Hence, the ADF tests support the earlier findings that both food and nonfood price inflation are persistent.

The next two tests check the null hypotheses that the first differences of food and nonfood inflation contain unit roots. Both the null hypotheses are rejected. Overall, these tests suggest that both food and nonfood inflation are I(1). Further, Johansen tests support the existence of a long-run relationship between food and nonfood inflation (Table 3B). We reject the null of no cointegration (trace statistic is greater than the 5 percent critical value).

Given the existence of a cointegrating relationship, we use two methodologies to examine the association between food and nonfood inflation. We ask the question—do changes in food inflation pass through to nonfood inflation? First given that Ordinary Least Squares (OLS) produces super-consistent estimates in the presence of cointegration, we use a simple linear regression of nonfood inflation on food inflation in order to estimate the pass-through. We find evidence of a statistically significant pass-through

18. We cannot reject the null hypothesis of the existence of a unit root in the nonfood inflation series only marginally at the 5 percent level.

TABLE 3A. Augmented Dickey–Fuller Test for Unit Root

	Test statistic	*5% critical value*	*p-value*	*Null*	*Outcome: 5% significance*
Food inflation	–3.035	–3.463	0.123	Unit root	Do not reject
Nonfood inflation	–3.496	–3.463	0.040	Unit root	Do not reject^
d.food inflation	–7.382	–1.664	0.000	Unit root	Reject
d.nonfood inflation	–6.445	–1.664	0.000	Unit root	Reject

TABLE 3B. Johansen Test for Cointegration between Food and Nonfood Inflation

	Trace statistic	*5% critical value*	*Null*	*Outcome*
Null: No cointegration	34.2754	15.41	No cointegration	Reject

Note: The tests are based on quarterly inflation rates (deseasonalized). Optimal lag length is selected based on the AIC Criterion. The null hypothesis for the ADF test is the presence of a unit root. Trend an intercept are included in the ADF regression to test for unit roots in levels; and only the intercept in the regression to test for unit root in differences. ^ denotes the null hypothesis is not rejected at the 1 percent level of significance.

from food price inflation to nonfood inflation at the 1 percent level. The magnitude moreover is economically significant. A one percentage point increase in food inflation is associated with a 0.2 percentage point increase in nonfood inflation (Table 4).

TABLE 4. Pass-through from Food to Nonfood

Dependent variable. Nonfood inflation	
Food inflation	0.195***
	[0.073]
N	91
r2	0.1

Note: This table reports the coefficients from an OLS regression of nonfood on food inflation, both measured as quarterly changes. Standard errors are denoted in parentheses. ***, **, and * denote statistical significance at 1, 5, and 10 percent, respectively.

Second, we estimate a Vector Error Correction Model (VECM), which allows for dynamics and explicitly includes an error correction term. The impulse responses of interest estimated from the VECM model are shown in Figure 12.[19] Two key findings emerge from the impulse responses. First, we

19. Standard errors for the impulse responses have to be estimated by bootstrapping methods; the codes are not available in standard software packages, and we are in process of developing our own codes.

FIGURE 12. Impulse Responses from a VECM Model

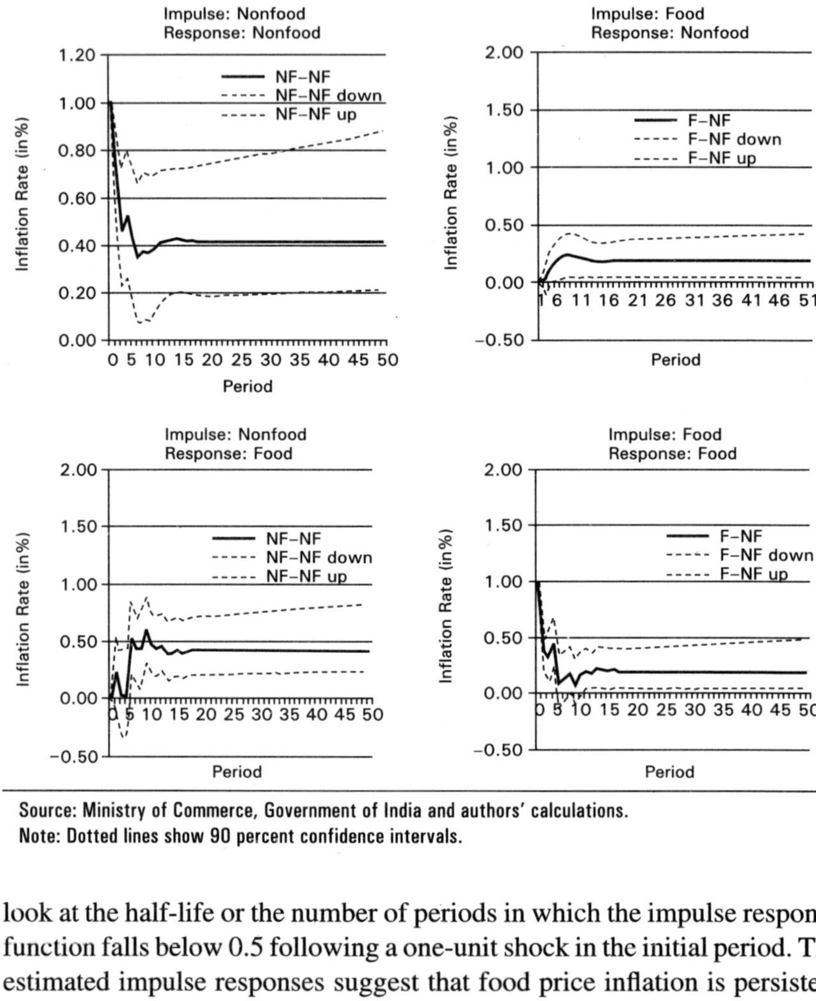

Source: Ministry of Commerce, Government of India and authors' calculations.
Note: Dotted lines show 90 percent confidence intervals.

look at the half-life or the number of periods in which the impulse response function falls below 0.5 following a one-unit shock in the initial period. The estimated impulse responses suggest that food price inflation is persistent with a half-life of three to six months. In comparison, the nonfood inflation is more persistent with a half-life of two quarters. The impulse responses for nonfood, however, are imprecisely estimated.

Furthermore, as in the simple OLS regressions we find some evidence for pass-through from food to nonfood inflation. The point estimate suggests that a one unit shock to food inflation (or a shock which results in a 0.2 percentage point long-run increase in the food inflation rate) increases the nonfood inflation by close to 0.1 percentage point in half a year. As inflationary expectations build up, the pass-through increases at the minimum by 0.05 percentage points in the long-run (based on the 90 percent confidence bands). Mechanisms other than the rise in general inflationary expectations could explain the pass-through as well. For example, a rise in food prices works as an income effect for net producers of food; and this could feed into increased demand for nonfood items. Finally, the tests for Granger causality provide some evidence for causality in both directions (from food to nonfood as well as the converse). Note that while Granger causality is a short run test, as discussed above we do find strong evidence for a long-run relationship between food and nonfood inflation rates.[20]

Relationship between Relative Price of Food and Overall Inflation

Finally, in this subsection, we revisit the classical question: do changes in the relative price of food feed into aggregate inflation? As discussed earlier, according to the Friedman view, this should not be the case. In order to make a first pass at this question, we look at the evolution of price of food relative to nonfood and aggregate WPI-based inflation in Figure 13. Although the

FIGURE 13. Relative Price of Food to Nonfood and Aggregate Inflation

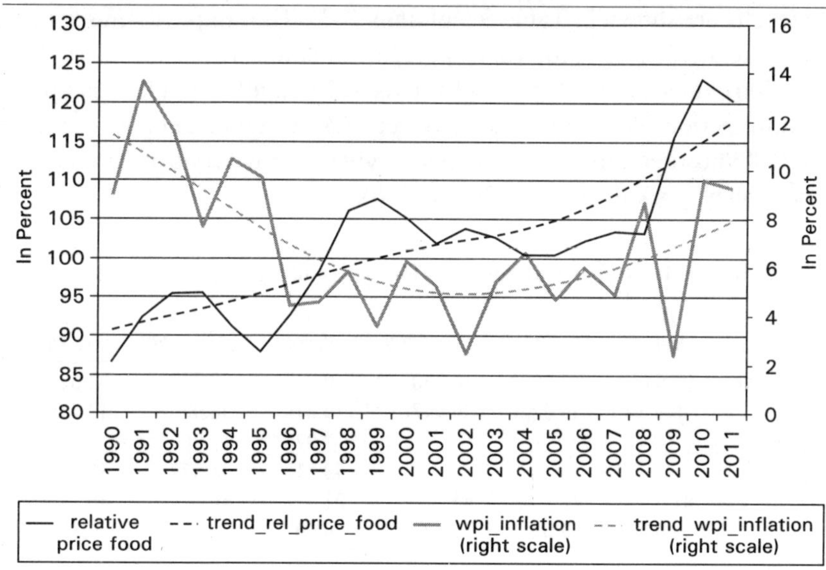

Sources: Ministry of Commerce and Industry, Government of India, and authors' calculations.
Notes: This figure shows the average inflation rate and the relative price of food to nonfood based on WPI. The trends are estimated based on a HP filter, with a smoothing parameter of 100.

20. For robustness, we repeated the VECM analysis in a three-variable system: (*a*) food price inflation, (*b*) nonfood inflation, and (*c*) overall inflation. The three variables exhibit a cointegrating relationship. The impulse response functions corroborate rest of the findings. We find evidence of pass-through from food to nonfood, and from food to aggregate inflation. We also find similar degree of persistence in food price inflation.

relative price of food has been trending upward since the early 1990s, there have been distinct subperiods (*a*) 1990–99, where the increase is likely to be explained by the economic reforms that improved the terms of trade for agriculture; (*b*) 2000–05, when the relative price of food in fact decreased; and (*c*) 2006 onwards, when the relative price of food has been increasing, but unlike in the earlier period, is driven primarily by high-value items like animal source foods, fruits and vegetables, and processed food—likely to be explained by the changing demand patterns.

There is a strong co-movement between the relative price of food and aggregate inflation rate in years 2000 and later. The relationship is stronger when we look at the long-term trends.

In order to analyze this relationship more formally, we use quarterly data and test for the existence of a long-term or cointegrating relationship between aggregate inflation (measured quarter-on-quarter), and the relative price of food. Based on strong evidence for cointegration, we estimate a simple OLS regression of aggregate inflation on the relative price of food. The results are shown in Table 5 (columns 1–3). The simple co-movement between relative price of food and aggregate inflation is not corroborated by regression evidence. We do not find a statistically significant relationship for the entire period. We do find some evidence for a significant pass-through post 2000; however, the results do not survive once we control for changes in the money stock (Table 5, columns 4–6).

TABLE 5. Relative Price of Food (to Nonfood) and Aggregate Inflation

	[1] Overall	*[2] Post-2000*	*[3] Pre-2000*	*[4] Overall*	*[5] Post-2000*	*[6] Pre-2000*
Relative price	−0.017	0.038*	−0.053***	−0.017	0.032	−0.052***
of food	[0.015]	[0.022]	[0.017]	[0.014]	[0.020]	[0.018]
Change in				7.978**	11.660***	1.442
money stock				[3.215]	[3.527]	[3.939]
N	91	46	45	91	46	45
r2	0.02	0.07	0.13	0.08	0.21	0.13

Note: The table reports the coefficients from OLS regressions based on quarterly data on WPI-based inflation (deaseasonalized) on relative price of food to nonfood. Columns (1)–(3) use data from 1988q1–2011Q2; columns (4)–(6) use data from 1988q1–2008q4. Money stock is given by M0. Standard errors are denoted in parentheses. ***, **, and * denote statistical significance at 1, 5, and 10 percent, respectively.

Implications

The results in this section establish the importance of food prices in overall inflation in India. An important policy question is whether central banks should focus on headline or core measures of inflation. In principle, core inflation measures should remove temporary noise (e.g., due to weather shocks that drive prices of food temporarily higher). As argued by Cecchetti (2007), measured over a longer horizon, headline and a well-designed measure of core inflation should coincide. However, given that the headline inflation has been consistently above core in the Indian case over the last two decades (the only exceptions being in 1994 and 1995 [Figure 14]), core inflation is likely to be a biased estimate of medium and long-term inflation.

Moreover, a prediction that comes out of the benchmark New Keynesian model with nominal price rigidities is that central banks should respond by tightening if there are second round effects of an increase in the relative commodity prices. The evidence we find for a significant association between relative price of food and aggregate inflation, and for pass-through from food

FIGURE 14. Inflation Rates: Headline and Core

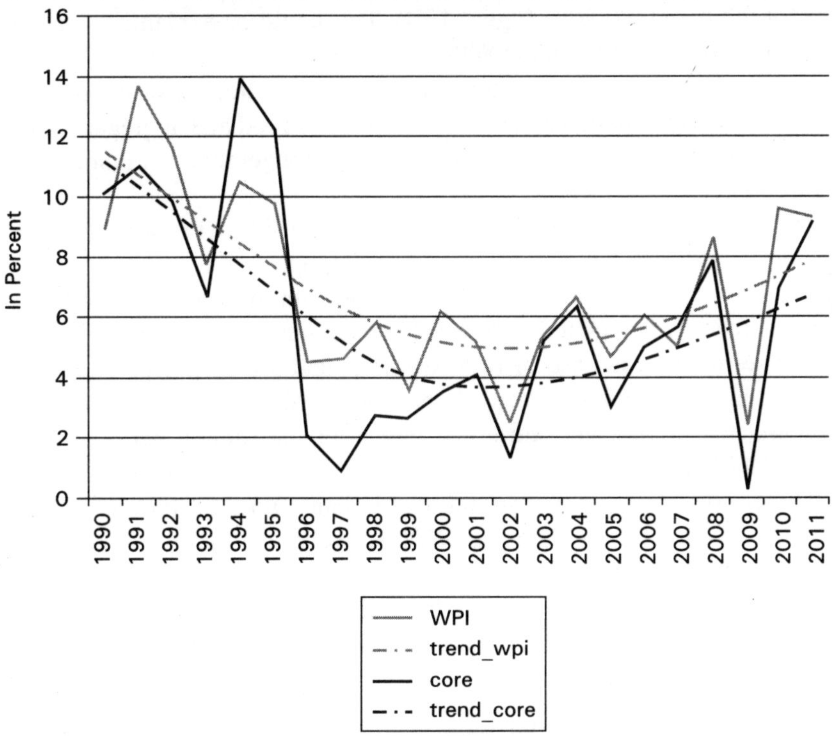

Sources: Ministry of Commerce and Industry, Government of India, and authors' calculations.

Notes: This figure shows the average y-o-y inflation rates based on WPI. Core inflation is defined over all commodities excluding food and fuel. The trends are estimated based on a HP filter, with a smoothing parameter of 100.

to nonfood inflation suggests that the central bank should not ignore food prices while making monetary policy decisions.

Overall, these results call for focusing on *headline* rather than *core* measures of inflation for policy purposes.

Can our analysis shed light on which measure of *headline* inflation should policymakers use? In order to answer this question, one of exercises we carry out is to evaluate the relationship between relative price of food and aggregate inflation with CPI-based indices. In order to conduct this analysis, we create sub-indices for food and nonfood (excluding fuel) using the CPI data. One surprising finding is that the food price inflation rate based on CPI tracks very closely the inflation rate based on the WPI (the correlation between the WPI-based and either of the CPI-based food price inflation rate [and also prices] is 0.9, Figure 15A). Ideally, one would expect the prices at the retail level to be significantly above the wholesale price due to markups along the value-chain.[21] In contrast, in the nonfood category, the CPI, and WPI-based inflation rates have very low correlation (particularly for the CPI for industrial workers, Figure 15B). This could be explained, e.g., by the inclusion of services in the WPI.

Figures 16A and 16B show that the CPI-based indices also support evidence for co-movement between relative price of food and aggregate inflation. Moreover, the relationship appears to be stronger than WPI (shown in Figure 12).

Similarly, repeating the exercise in Table 2 using CPI-based indices, the results suggest a much stronger pass-through, with the effect being strongest for rural workers. A one percentage point increase in the relative price of food is associated with a 0.12 percentage point increase in the aggregate inflation in case of rural workers, and a 0.06 percentage point increase in case of industrial workers. These effects are statistically and economically significant, and are stronger for the sample period before 2000 (Table 6).[22]

These results suggest that an index of inflation relevant for consumers, CPI may be more appropriate than WPI. However, CPI in its current form has very limited coverage and cannot be used for disaggregate sector-level analysis and, is not used in practice by Indian policymakers. A case could therefore be made for building a more comprehensive CPI.

21. Basu (2011) also finds that restricted to a common set of commodities, the WPI and CPI based indices are highly correlated.

22. Table 6 only shows the regressions without controlling for money stock. The results remain remarkably similar when we control for changes in money stock.

FIGURE 15A. Food Price Inflation Rate: CPI and WPI

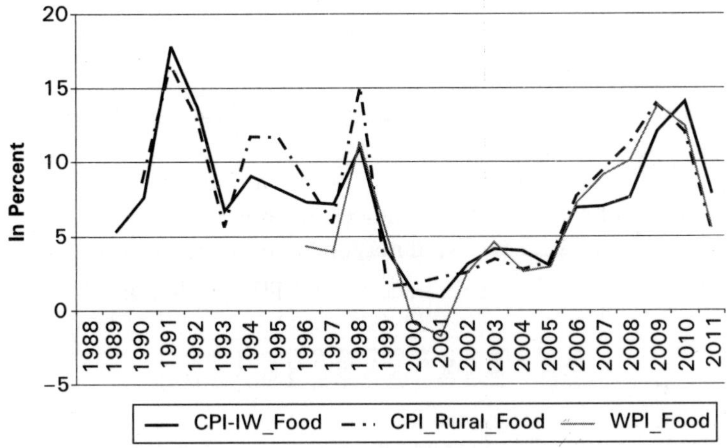

Source: Ministry of Commerce and Industry, Government of India.

Notes: This figure shows the average y-o-y food inflation rates.The WPI figure for 2011 is based on Jan–May; for CPI, from Jan–April. CPI-IW, CPI-AL, CPI-RL and CPI-UNME are the CPI for industrial, agriculture, rural, and urban nonmanual employees, respectively.

FIGURE 15B. Nonfood Inflation Rate: WPI and CPI

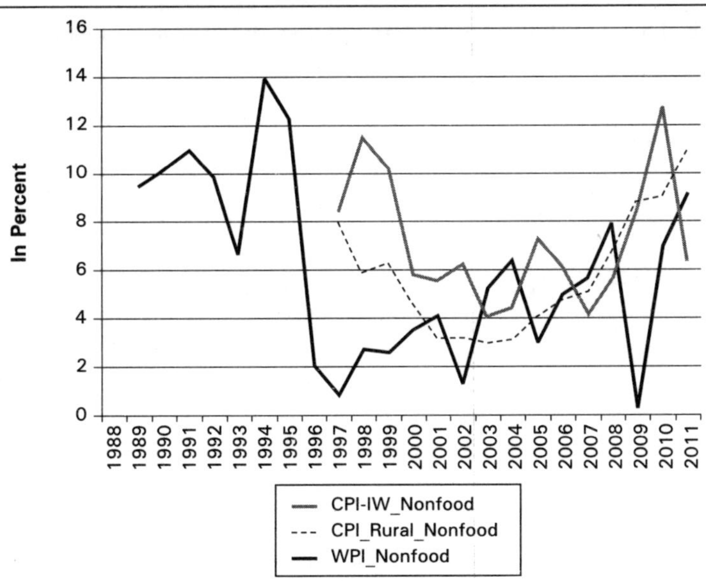

Source: Ministry of Commerce and Industry, Government of India.

Notes: This figure shows the average y-o-y nonfood (excluding) inflation rates. The WPI figure for 2011 is based on Jan–May; for CPI, from Jan–April. CPI-IW, CPI-AL, CPI-RL, and CPI-UNME are the CPI for industrial, agriculture, rural, and urban nonmanual employees, respectively.

FIGURE 16A. Relative Price of Food and Overall Inflation (CPI–Industrial Workers)

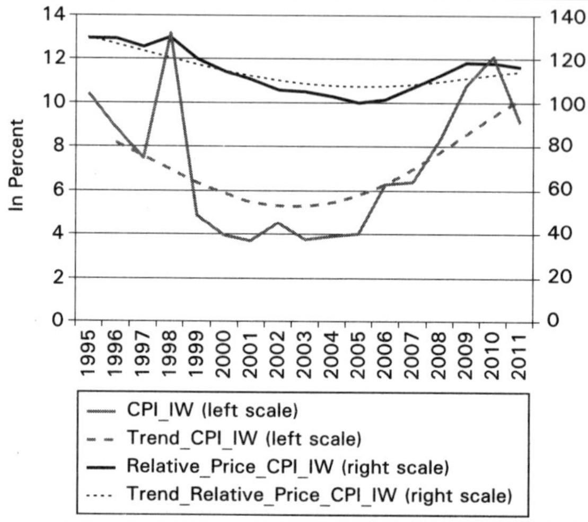

Sources: Ministry of Commerce and Industry, Government of India, and authors' calculations.

Notes: This figure shows the average inflation rate and the relative price of food to nonfood based on CPI for industrial workers. The trends are estimated based on a HP filter, with a smoothing parameter of 100.

FIGURE 16B. Relative Food Prices and Overall Inflation (CPI–Rural Workers)

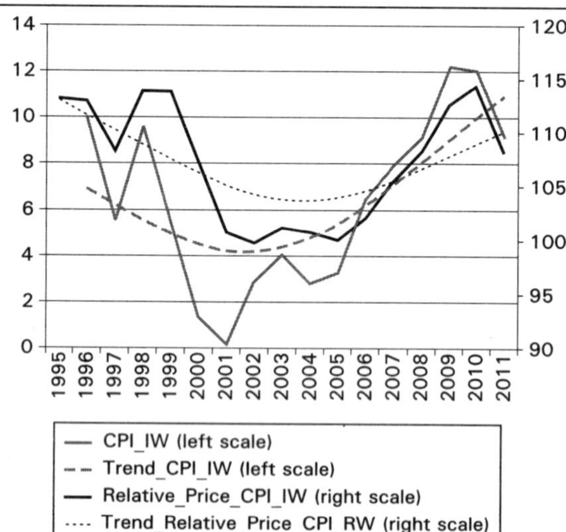

Sources: Ministry of Commerce and Industry, Government of India, and authors' calculations.

Notes: This figure shows the average inflation rate and the relative price of food to nonfood based on CPI for rural workers. The trends are estimated based on a HP filter, with a smoothing parameter of 100.

TABLE 6. Relative Price of Food (to Nonfood) and Aggregate Inflation based on CPI

	CPI-IW			CPI-Rural		
	[1]	[2]	[3]	[4]	[5]	[6]
	Overall	Post-2000	Pre-2000	Overall	Post-2000	Pre-2000
Relative price of food	0.057*** [0.016]	0.084*** [0.021]	0.279*** [0.072]	0.121*** [0.026]	0.143*** [0.033]	0.247** [0.085]
N	65	45	20	63	47	16
r2	0.18	0.29	0.36	0.23	0.34	0.20

Note: These are based on OLS regressions based on quarterly data of aggregate CPI-based inflation on relative price of food to nonfood. Standard errors are denoted in parentheses. ***, **, and * denote statistical significance at 1, 5, and 10 percent, respectively.

Which Commodities Drive Food Price Inflation?

There are two main factors which determine the contribution of different commodities to food price inflation: (*a*) the weight of each commodity in the overall food basket and (*b*) change in prices of these commodities. To begin with, we look at broad trends in inflation of primary and manufactured food items. From Figure 2, we know that primary food items constitute a bigger weight in the food basket. At the same time, the inflation rate for primary food items is typically higher than that for manufactured items (Figure 17). For example, in 2010, the difference was as high as 10 percentage points. Moreover, the long-term trend in the inflation for primary products is always above that for manufactured food items. Hence, given the high weights, and high inflation rates, we can expect primary commodities to be contributing to a larger extent to the overall food inflation.

In order to dig deeper into the specific commodities which contributed to food price inflation, we look at all the 112 commodities in the 2004–05 food basket and use a simple methodology to quantify their contributions.

Recall in Equation (2), we defined the food price index using a Laspeyres formula. Taking first differences of (2), and dividing by I^F_{t-1},

$$\frac{\Delta I^F_t}{I^F_{t-1}} = \sum_i \frac{W^F_i}{\sum_i W^F_i} * \frac{\Delta I^F_{it}}{I^F_{it-1}} * \frac{\Delta I^F_{it-1}}{I^F_{t-1}}$$

Or,

$$\pi^F_t = \sum_i \frac{W^F_i}{\sum_i W^F_i} * \pi^F_{it} * \frac{I^F_{it-1}}{I^F_{t-1}} \quad (3)$$

FIGURE 17. Inflation Rate: Food. Primary and Manufactured Products

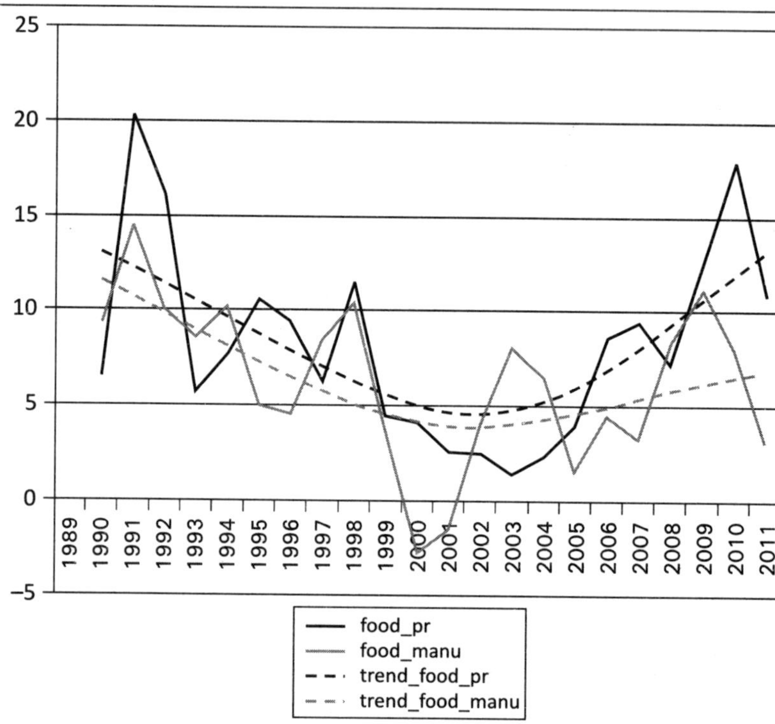

Sources: Ministry of Commerce and Industry, Government of India, and authors' calculations.

Notes: This figure shows the average y-o-y inflation rates based on WPI. The trends are based on a HP filter with a smoothing parameter of 100.

Where W_i^F is the weight for i in the food basket in the base period. π_t^F denotes the aggregate food inflation rate and π_{it}^F denotes the inflation rate for commodity i at time t. Hence, the contribution of item i in explaining food inflation is given by

$$C_{it}^F = \frac{W_i^F}{\sum_i W_i^F} * \pi_{it}^F * \frac{I_{it-1}^F}{I_{t-1}^F} \tag{4}$$

C_{it}^F is a product of three factors—(a) the share of commodity i in the food basket, (b) the inflation rate of i, and (c) the ratio of the price index of i to the overall food price index in the previous period. Hence, commodities with higher weights as well as high inflation are the natural candidates for being among the bigger contributors to the overall food inflation. However, commodities with very high weights in the food basket could also contribute significantly even though the rise in their prices is not very significant. The converse could be true as well—there could be commodities with relatively low weights which experience a sharp increase in prices, and could contribute to overall food inflation.

Based on this logic, we classify all the commodities in the food baskets into four bins: (*a*) high inflation and high weight, (*b*) high inflation and low weight, (*c*) low inflation and high weight, (*d*) low inflation and low weight. High and low are defined as above and below the median of inflation and weight respectively. The high-weight, high-inflation bin for selected years—1998, 2008, 2009, and 2010 is shown in Table 7.[23] The commodities falling in this bin are the most likely candidates for the biggest contributors to food inflation. Though there is some variation in the list of commodities, broadly the high inflation and high weight commodities can be classified into four groups: (*a*) animal source food—which includes milk, fish, and chicken; (*b*) fruits and vegetables, e.g., mangoes, cauliflower, brinjal, etc.; (*c*) staples, e.g., rice and wheat; and (*d*) processed food—including sugar, edible oils, etc.

In what follows, from Equation (4), we compute the LHS variable of Equation (4) for each of the 112 commodities in the food basket. We first take the annual averages of the individual commodity indices as well as of the overall food index, and then calculate the respective inflation rates. We use the weights in the 2004–05 basket for the calculations. Finally, as a check, we make sure that the following identity holds in the data.

$$\sum_{i} C_{it}^{F} = \pi_{t}^{F}$$

The top 50 contributors and their contributions for some selected years—1998, 2008, 2009, and 2010 are shown in Table 8. The top five contributors in each of these years are also shown in Figure 18. The contributors to food inflation are typically concentrated in a few commodities. The total contribution of the top five to the overall food inflation is large—46 percent in 1998, 44 percent in 2008, 59 percent in 2009, and 51 percent in 2010. The top contributors are typically the commodities in the high-inflation, high-weight bin. They include milk and fish in the category of animal source food; onions, potatoes, cauliflower, and mangoes in fruits and vegetables; sugar and edible oils (mustard and rapeseed oil, and vanaspati) in processed food; and rice and wheat in cereals.

23. All the four bins for some selected years are shown in Table A1.

TABLE 7. List of High-Inflation and High Weight Commodities: 1998, 2009, and 2010

	Animal source products	*Fruits and vegetables*	*Staples (cereals/pulses)*	*Processed food*
2010	Beef and Buffalo Meat (41.6,0.5)	Mangoes (34.9,2.7)	Arhar (11.2,0.6)	Ghee (17,0.9)
	Marine Fish (40.5,3)	Cauliflower (24.1,1)		Groundnut Oil (13.5,1.3)
	Inland Fish (37.8,2.4)	Oranges (22.5,0.5)		Processed Prawn (16.4,0.5)
	Milk (24.6,13.3)	Papaya (21.8,0.4)		Sugar (14.7,7.1)
	Poultry Chicken (23.7,1.7)	Cabbage (21.7,0.8)		Atta Wheat Flour (10.4,1.6)
	Egg (21.7,0.8)	Brinjal (20,1.2)		Betelnuts, Arecanut (9.6,0.4)
	Mutton (11.9,1.4)	Banana (19.2,1.4)		Biscuits (9.3,1.4)
		Okra (13.3,0.5)		Maida (9.2,1.9)
		Tomatoes (13,1.1)		Cattle Feed Gola (11.3,0.5)
		Onions (11.9,0.7)		
2009	Inland Fish (34.2,2.4)	Potatoes (70.9,0.8)	Arhar (39.2,0.6)	Molases (53,0.5)
	Marine Fish (8.8,3)	Onions (36.9,0.7)	Bajra (20.8,0.5)	Sugar (47.7,7.1)
	Mutton (20.5,1.4)	Cashew Kernel (29.8,1.6)	Rice (13.7,7.4)	Unblended Tea Leaf (22.3,0.7)
	Milk (14.3,13.3)	Apples (27.2,0.4)	Maize (11,0.9)	Cotton Seed (21.3,0.5)

		Cashwenuts (19.3,0.7) Papaya (10.7,0.4) Green Peas (10.5,0.5)	Wheat (10.6,4.6)	Dry Chillies (19.3,0.7) Ghee (17.3,0.9) Blended Tea Leaf (15.2,1.1) Cattle Feed Gola (14.7,0.5) Blended Tea Dust (12.1,0.6) Other Oil Cakes (10.8,0.8)
1998	Inland Fish (10.5,2.7) Marine Fish (8,3.4) Milk (8.5,15.1)	Onions (186.2,0.8) Potatoes (152.8,0.9) Mangoes (75.2,3) Cauliflower (73.1,1.1) Cabbage (50.7,0.9) Oranges (45.3,0.6) Tomatoes (38.8,1.2) Brinjal (25.7,1.4) Okra (23.1,0.6) Cashwenuts (7.8,0.8)	Arhar (29.5,0.6)	Mustard and Rapeseed Oil (42.3,2.1) Rice Bran Oil (34.9,0.9) Soyabean Oil (34.9,1.8) Unblended Tea Leaf (23.3,0.8) Cotton Seed Oil (20.2,1.2) Groundnut Oil (19.8,1.4) Sunflower Oil (16.1,0.8) Ghee (9.6,1) Biscuits (7.9,1.6)

Notes: The first and second figure in parentheses indicate the inflation rate and weight in the food basket respectively. Two additional high-inflation, high-weight commodities not falling under any of these categories are Tea (27.3,0.5) and Dry Chillies (31.8,0.7).

TABLE 8. Contributions of Top 50 Contributors to Food Inflation (in percent). Selected Years

Rank	*1998*		*2003*		*2008*		*2009*		*2010*	
1	Mangoes	12.8	Mustard/Rapeseed Oil	13.6	Rice	12.6	Sugar	23.4	Milk	22.2
2	Onions	10.6	Cauliflower	13.1	Mangoes	11.2	Milk	15.6	Marine Fish	9.0
3	Milk	8.1	Marine Fish	10.2	Milk	10.1	Rice	9.3	Sugar	7.7
4	Potatoes	7.2	Mangoes	9.8	Wheat	5.4	Inland Fish	5.5	Mangoes	6.5
5	Cauliflower	6.9	Vanaspati	8.7	Sugar	4.4	Wheat	4.8	Inland Fish	5.9
6	Mustard/Rapeseed Oil	5.5	Soyabean Oil	7.2	Marine Fish	4.2	Potatoes	4.6	Rice	4.2
7	Tomatoes	4.9	Milk	6.8	Tomatoes	3.3	Cashew Kernel	3.1	Turmeric	3.2
8	Soyabean Oil	4.1	Groundnut Oil	6.1	Mustard/Rapeseed Oil	2.9	Onions	2.7	Wheat	2.8
9	Cabbage	3.9	Rice	5.7	Vanaspati	2.9	Mutton	2.7	Garlic	2.4
10	Rice	3.5	Atta Wheat Flour	4.8	Brinjal	2.7	Marine Fish	2.5	Poultry Chicken	2.2
11	Brinjal	2.9	Maida	4.6	Others	2.5	Arhar	2.1	Banana	1.8
12	Marine Fish	2.4	Coconut	4.2	Cauliflower	2.3	Dry Chillies	1.4	Brinjal	1.6
13	Dry Chillies	2.3	Brinjal	4.0	Soyabean Oil	2.2	Urad	1.4	Moong	1.6
14	Sugar	2.3	Unblended Tea Leaf	3.7	Cotton Seed Oil	2.2	Blended Tea Leaf	1.4	Urad	1.5
15	Cotton Seed Oil	1.9	Cotton Seed	3.5	Mutton	2.1	Gur	1.4	Cauliflower	1.4
16	Groundnut Oil	1.9	Rice Bran Oil	3.5	Tea	1.8	Molases	1.4	Mutton	1.4
17	Rice Bran Oil	1.8	Dry Chillies	2.8	Blended Tea Leaf	1.7	Turmeric	1.4	Cabbage	1.3
18	Wheat	1.7	Banana	2.7	Banana	1.5	Moong	1.3	Beef and Buffalo Meat	1.3
19	Inland Fish	1.5	Wheat	2.7	Palm Oil	1.5	Unblended Tea Leaf	1.3	Atta Wheat Flour	1.2
20	Unblended Tea Leaf	1.5	Powder Milk	2.6	Corriander	1.4	Ghee	1.2	Guava	1.1
21	Oranges	1.4	Turmeric	2.4	Papaya	1.2	Tea	1.2	Maida	1.1
22	Arhar	1.4	Copra Oil	2.3	Coconut	1.1	Tapioca	1.2	Egg	1.1
23	Bagasse	1.2	Onions	2.3	Rice Bran Oil	1.0	Apples	1.0	Tomatoes	1.1
24	Okra	1.1	Cabbage	2.1	Groundnut	1.0	Cashwenuts	1.0	Groundnut Oil	1.0
25	Poultry Chicken	1.1	Blended Tea Leaf	2.0	Arhar	1.0	Maize	0.9	Ghee	1.0

26	Biscuits	1.1	Tapioca	1.7	Masur	1.0	Bajra	0.9	Onions	0.9
27	Turmeric	1.0	Arhar	1.3	Atta Wheat Flour	0.9	Cotton Seed	0.9	Oranges	0.9
28	Tea	1.0	Ghee	1.2	Ghee	0.8	Atta Wheat Flour	0.8	Tapioca	0.9
29	Sunflower Oil	0.9	Oranges	1.1	Rice Bran Extraction	0.8	Banana	0.8	Vanaspati	0.8
30	Black Pepper	0.9	Mutton	1.0	Unblended Tea Leaf	0.8	Others	0.8	Biscuits	0.7
31	Garlic	0.8	Canned Fish	0.9	Maida	0.8	Khandsari	0.8	Rice Bran Extraction	0.7
32	Vanaspati	0.8	Green Peas	0.9	Coffee	0.8	Masur	0.7	Papaya	0.7
33	Grapes	0.7	Gingelly Oil	0.9	Oranges	0.8	Unblended Tea Dust	0.7	Blended Tea Leaf	0.6
34	Ghee	0.7	Rawa Sooji	0.9	Cattle Feed Gola	0.8	Poultry Chicken	0.7	Arhar	0.6
35	Papaya	0.5	Okra	0.9	Ginger	0.7	Cattle Feed Gola	0.6	Litchi	0.6
36	Mutton	0.5	Bagasse	0.8	Unblended Tea Dust	0.7	Maida	0.6	Cashew Kernel	0.6
37	Cashwenuts	0.5	Sunflower Oil	0.7	Excluding Oil	0.7	Egg	0.5	Cardamoms	0.5
38	Maida	0.4	Groundnut	0.7	Guava	0.6	Blended Tea Dust	0.5	Maize	0.5
39	Moong	0.4	Coffee	0.7	Biscuits	0.6	Groundnut	0.5	Powder Milk	0.5
40	Sapota	0.3	Bajra	0.7	Salt	0.6	Coffee	0.5	Okra	0.5
41	Gur	0.3	Cattle Feed Gola	0.7	Sunflower Oil	0.6	Biscuits	0.4	Cattle Feed Gola	0.4
42	Salt	0.3	Mustard	0.6	Gur	0.6	Oranges	0.4	Processed Prawn	0.4
43	Blended Tea Leaf	0.3	Maize	0.6	Rice Bran	0.6	Papaya	0.4	Pork	0.4
44	Gingelly Oil	0.3	Guava	0.5	Gingelly Oil	0.6	Rice Bran Extraction	0.4	Gur	0.4
45	Khandsari	0.3	Cashwenuts	0.5	Mustard	0.5	Garlic	0.4	Jowar	0.4
46	Gram	0.2	Dry Ginger	0.4	Tapioca	0.5	Green Peas	0.4	Wheat Bran	0.4
47	Masur	0.2	Apples	0.4	Maize	0.5	Mustard	0.4	Butter	0.3
48	Butter	0.2	Corriander	0.4	Turmeric	0.5	Vegetables Seeds	0.4	Bajra	0.3
49	Mustard	0.2	Papaya	0.4	Groundnut Oil	0.5	Powder Milk	0.3	Bagasse	0.3
50	Apples	0.2	Ragi	0.4	Grapes	0.5	Sapota	0.3	Rawa Sooji	0.3

FIGURE 18. Contribution of Leading Commodities to Food Inflation

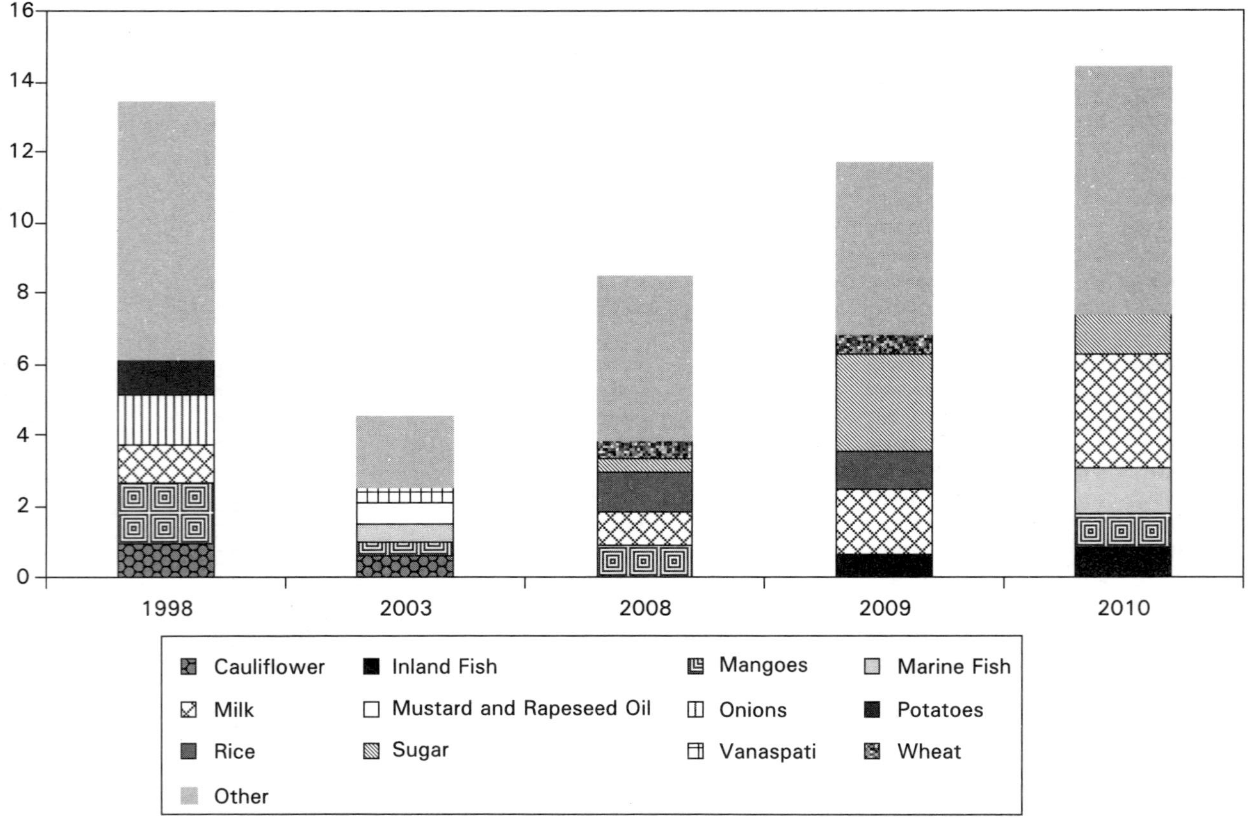

Sources: Ministry of Commerce and Industy, Government of India, and authors' calculations.

There has been some variation in the top group as well. For example, in 2010, animal source food was the leading group, both animal source foods and processed foods were equally important in 2009; processed food was leading in 2008; while in 1998, fruits and vegetables dominated the inflation. The leading individual items contributing to food inflation has also been changing—milk in 2010, sugar in 2009, rice in 2008, and mangoes in 1998.[24] Milk has been important over the entire time period and its significance has increased. It explained about 8 percent of food inflation in 1998, and its importance has increased three times by 2010.

What Factors Explain the Rise in Food Price Inflation? Case Studies of Specific Commodities

In this section, we examine in detail the possible factors that could explain the inflation patterns for commodities which we identified as primary contributors to food inflation during the last two decades. These factors can be classified into long-term factors and those that are more important in the shorter term.

The long-term factors include both the structural factors as well as long-term government policies. In particular, we consider the following: (*a*) demand-side factors, e.g., those related to changing structure of demand away from cereals toward high-value items like livestock products. This can be attributed to rising incomes and changing lifestyles (e.g., from urbanization);[25] (*b*) supply-side factors, e.g., changes in production and productivity. On the supply side, the performance of the agricultural sector in India has been abysmal. The sector has managed an average growth rate of merely 3 percent and with high volatility in the last 20 years. In particular, India has now lower yields per hectare of cereals than most comparable countries including Sri Lanka, China, Bangladesh, and Pakistan;[26] (*c*) long-term policies, e.g., the cereal-centric focus of the government through a system of producer support prices and maintenance of grain reserves. The three

24. Since mango is a seasonal fruit, the inflation figures are based on the few months, on which data is available.

25. For example, Mittal (2006) estimates the expenditure elasticities in 1999 for different food groups in India based on rounds in the NSS. She finds the highest elasticities for animal source food (e.g., milk), followed by processed food (e.g., sugar) and fruits and vegetables.

26. For example, in rice, data from FAO for 2008 suggests that yields in China at 6.5 tons per hectare are almost double that of India at 3.4 tons per hectare. Even the yields in Bangladesh are higher at 3.9 tons per hectare.

long-term structural factors contribute towards built-in inherent inflationary pressures, making the system vulnerable to price increases from short-term shocks.

In addition to longer-term structural factors and policies, there can also be short-term contributors to inflation. These include (*a*) short-term shocks, e.g., negative shocks like droughts, floods, etc., as well as positive income shocks; (*b*) domestic policy interventions, e.g., revision to minimum support prices (MSPs); (*c*) trade policy responses, e.g., easing of export restrictions; and (*d*) movements in international commodity prices.

In the previous section, we identified four groups as leading contributors to food inflation, namely, animal source food, fruits and vegetables, processed food, and cereals. We will analyze the trends in inflation for some selected items under these groups which we identified to be important.

Animal Source Food—Milk

Milk was the most important contributor to food inflation in 2010. Moreover, its contribution has steadily increased over time. Figure 19 shows the evolution of the inflation rate and production of milk over time. Three stylized facts emerge from the figure: (*a*) The inflation rate for milk shows an increasing trend since 2005, with particularly sharp increases in 2009 and 2010. (*b*) Based on data up to May, there was a reversal in inflation in 2011. As discussed earlier, we classify the factors potentially responsible for explaining the inflationary patterns in milk among long-term and short-term factors.

LONG-TERM FACTORS There has been a sustained increase in the production of milk over time. Milk production has nearly doubled between 1995 and 2010. Moreover, the rate of increase in production has risen from 3.7 percent prior to 2005 to 4.2 percent thereafter. In spite of being the largest producer of milk in the world, estimates show the productivity to be quite low. Annual milk yield per dairy animal was about one-tenth of that achieved in the USA and about one-fifth of the yield of a New Zealand dairy cow (FAO, 2003). Also, the milk availability was quite low at 241 grams per capita per day (Economic Survey, 2005–06).

According to Rakotoarisoa and Gulati (2006), Indian dairy policies have always been protecting dairy farmers grouped in cooperatives from low-priced dairy imports. The policies have included both domestic support and, in the past, high trade protection measures. The domestic support has been in the form of subsidies under the Operation Flood (OF) Program of the government which were part of the plan expenditure. From the point of view of the effect on prices, the OF program would be important not only in

FIGURE 19. Milk

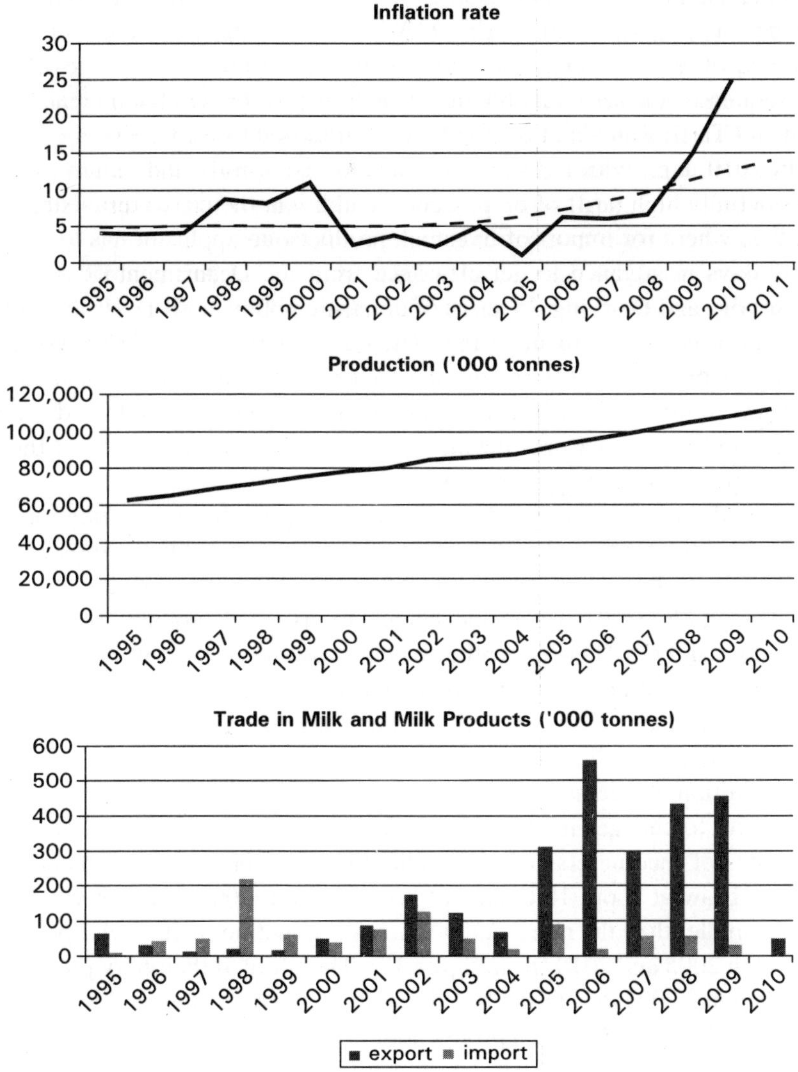

Sources: Ministry of Commerce and Indsutry, Government of India and FAOstat.
Notes: The dotted line in the first panel shows the HP filtered trend.

terms of increasing overall supply, but also in smoothing price differentials across time and space as it aimed to create a national milk grid.

The long-term trade policy of the government in milk and milk products has also played a role in sustaining inflationary pressures. Milk traditionally has faced high trade barriers. The trade protection before 1990 was mainly in the form of quota and canalization where all imports were controlled by National Dairy Development Board (NDDB). However, the import of milk powder, as per one of the earlier GATT agreements, was being allowed at a rate of zero percent. There was a spurt in imports of milk powders. Subsequently, the zero-duty bound rate in milk powder was finally renegotiated and Tariff Rate Quotas (TRQs) were imposed toward 2000.

In 2010, e.g., with a 3 percent education cess, milk and cream face a substantially high tariff of 68 percent. In addition, there are administrative hurdles, where for import of livestock products, an applicant has to apply thirty days in advance to get clearance from the Department of Animal Husbandry and Dairying. In terms of domestic policy, until 1991, the dairy industry was subject to licensing. The dairy industry was progressively de-"canalized" after 1991 and private sectors including multinational companies with milk processing and manufacturing plants were allowed.[27] All these supply-side measures meant that milk production increased in India but the growth turned out to be insufficient.

Thus the most important factor contributing to the rising inflation in milk, given the continued increase in production, is probably the substantial increase in its demand over time. Based on the different rounds in the NSS, Gandhi and Zhou (2010) show the demand for livestock products to be going up significantly.[28] Their expenditure share rose to 21 percent in 2004/05, being the second largest after cereals. Furthermore, they show that within livestock products, milk and milk products have the largest share at nearly three quarters in both rural and urban areas.

In addition, the demand for milk and milk products also have the greatest sensitivity to changes in income (proxied by expenditure). Based on their estimates of income elasticity, Gandhi and Zhou project the demand for milk to grow at about 10.6 percent per year from 2004 to 2005, which is much smaller than the rate of growth in its production of about 4.2 percent (between 2005 and 2010 from Figure 19). Therefore, inflationary pressures in milk since 2005 can be attributed mainly to the rising demand which has outpaced the increases in its production.

27. In 1992, because of concerns raised about excessive capacity of milk products and sale of adulterated/contaminated milk, limited controls were brought back through the Milk and Milk Products Order (MMPO), see Rakotoarisoa and Gulati (2006). MMPO from June 1992, reintroduced registration for milk processing units and was thus an entry barrier for private sector. In March 2002, the government made some important amendments so that the MMPO would basically restrict itself to regulating food safety, quality, sanitary and hygiene conditions of all registered units (see Jha, 2004).

28. Gokarn (2011) also presents evidence for rising demand for proteins, both animal (e.g., milk) and plant (e.g., pulses) based due to rising incomes and changing lifestyles.

Short-term Factors In addition to longer-term structural factors and policies, we also identify some key short-term shocks—both positive and negative, which could have contributed to the observed patterns of inflation in milk.

Two key developments since 2005/06 which constitute positive income shocks to households are worth mentioning. First, the Mahatma Gandhi National Rural Employment Guarantee Scheme (MGNREGS) was introduced with the first phase starting early 2006. Second, the sixth pay commission was implemented starting early 2009. Both these schemes brought significant increases in the disposable incomes of households. Combined with the high income elasticity of milk, these developments could contribute toward explaining the rising trend in its inflation since 2005, and the spikes in 2009 and 2010.

Furthermore, the droughts in 2009 in North India that hiked the price of fodder and raised the cost of milk production could also have reinforced the inflationary pressures from the input side.

Finally the reversal in the inflation rate in 2011 (based on data till May) can be due to two main reasons. First, the improved supply of fodder possibly reduced the costs of milk production and eased inflationary pressures. Second, the recent trade policy responses by the government could have played a role as well. The government allowed duty-free import of 30,000 tons of skimmed milk powder and 15,000 tons of butter oil in February 2011, which could have eased the supply pressures. In addition, the government imposed an export ban on milk powder and casein at the same time. Since both exports and imports of milk and milk products constitute a very small share of production (less than 1 percent), these trade measures could at most play a marginal role toward easing inflationary pressures in milk. Since the drop in the inflation rate to single digits is fairly recent, it remains to be seen how far these measures remain successful in taming inflation.

Sugar

Two distinct facts stand out from Figure 20A: (*a*) There have been three peaks in the inflation rate during the last two decades—15.2 percent in 1997, 16.6 percent in 2004, and the highest peak of 47.0 percent in 2009. Excluding these peaks, the trend in sugar prices has actually been deflationary. (*b*) The pattern on inflation closely tracks the pattern of sugarcane (and sugar) production.

Long-term Factors Sugar has a reasonably high income elasticity of demand. Among the elasticities estimated by Mittal (2006), with an elasticity

FIGURE 20A. Sugar

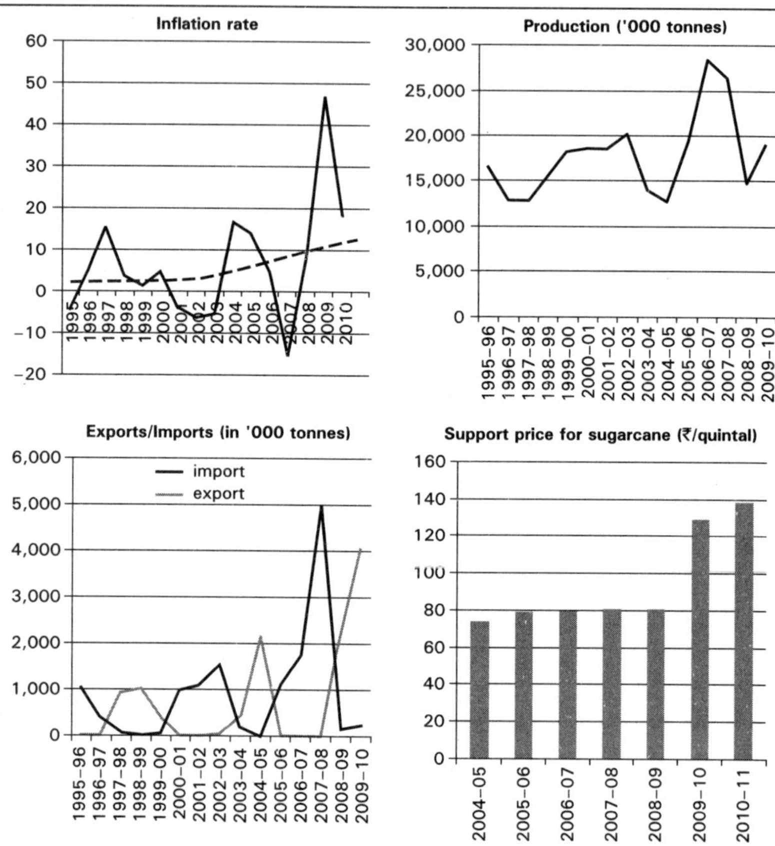

Source: Ministry of Commerce and Industry, Government of India, FAOstat and Indiastat.

of 0.83, it ranks second after animal source foods. Not only the direct but also the derived demand for sugar from the processing industry (such as soft drinks, biscuits, and confectionaries) has been rising. Hence, similar to animal source foods, the rising demand accompanying economic growth is a long-term structural factor contributing to sugar inflation. However, one key difference in the case of sugar is that, in spite of the long-term trend of rising demand, the demand-supply gap has not been persistent. In fact, in years of good harvest of sugarcane, there tends to be a match between total production and domestic demand with surplus if any (over accumulation of stocks) being allowed to be exported. In a year like 2006–07, India harvested 355 million tons of sugarcane converting to 28 million tons of sugar, while the demand for sugar was estimated to be only 20.8 million tons

(Edelweiss, 2010). However, in a bad harvest year such as 2008–09, when the production of sugar collapsed to 14.5 million tons; even assuming no demand growth, there is a significant demand-supply gap.

Long-term policies (both domestic and foreign) have played an important role in building up inherent inflationary pressures primarily because the sugar sector has been among the most heavily regulated industries in the country.

The complex system of support prices for different crops implies that sugarcane production depends significantly on policy measures in competing sectors as well, e.g., in cereals. The central government establishes a MSP for sugarcane on the basis of recommendations by the Commission for Agricultural Costs and Prices (CACP) and after consulting state governments and associations of the sugar industry and cane growers. Since 2009, the government has instituted a new system of fair and remunerative prices (FRP) that would link cane prices with sugar price realization by the mills. The FRP is used by the cane producing states of Maharashtra, Karnataka, Andhra Pradesh, and Bihar. Several state governments further augment the MSP/FRP, typically by 20–25 percent, due to political compulsions rather than market pricing. In addition, other big cane producing states such as Uttar Pradesh (UP), Uttarakhand, Punjab, Haryana, and Tamil Nadu declare their own State Advised Price (SAP) for cane. The SAP is usually higher than the FRP. For the 2010–11 sugar season, the FRP was ₹139 per quintal while the SAP in UP, the second biggest sugar producing state, was ₹205 per quintal. In addition to mandates for FRP and SAP, the government has also been levying a fee on sugar mills (currently ₹240 [$5.4] per ton of sugar) to fund a Sugarcane Development Fund (SDF).

The government interventions have not been restricted to sugarcane sector per se, but encompass regulations related to marketing, stocking, and futures trading as well. In marketing, the government has followed a policy of partial control and dual pricing. The sugar mills are mandated to supply 10 percent of their production to the government as levy sugar at below-market prices, which the government then channelizes to the consumers through the Public Distribution System (PDS) to the BPL families. Mills are allowed to sell the balance of their production at market prices. However, the release of free-sale sugar and levy-sugar is administered by the government through periodic quotas (for three months), designed to maintain price stability in the market.

In addition, the government can also impose stocking limits under the Essential Commodities Act. These limits are set as a proportion of requirements for particular time periods. Prior to 2001, the government did not allow any futures trading in sugar. Since May 2001 the rules were relaxed and three national exchanges were given permission to engage in futures trading. Finally, in terms of trade policy, the sugar sector has traditionally been closed, with tariffs currently at a level of 60 percent. In general, there has been a ban on imports of raw sugar.

Overall, a plethora of policy interventions on a regular basis could have long-term effects in terms of the incentives that they generate (affecting production, productivity, and marketing behavior) and efficiency of the system.

SHORT-TERM FACTORS The short-term factors contributing to the inflation pattern in case of sugar are related to (*a*) supply shocks, (*b*) changes in support prices, policies related to stocking and futures trading, and (*c*) temporary trade policies (Figure 20B). Broadly, sugar prices inversely follow the production of sugar which follows the production of sugarcane in the country. In 2008–09 when the production of sugarcane declined by almost a quarter, that of sugar declined by 45 percent and prices spiked. Once sugarcane production recovered, the upward pressure on prices eased.

On domestic policies, the support prices (FRP) of sugarcane were raised in 2009–10 which could have added to the rise in the prices of sugar. Since sugar mills in some states were required to pay the SAP (higher than FRP) to sugarcane farmers irrespective of the market price of sugar; it could have further added to inflationary pressures. Softening sugar prices in 2010 coupled with the anticipation of a large cane crop discouraged the sugar mills to pay higher cane prices. Nonetheless, cane prices received by farmers remained high because of government policies to not adjust the FRP or SAP downward implying that sugar prices stayed firm for some time.

On the marketing side, first, in May 2009, the government suspended futures trading in sugar until September 2010. Due to improvements in domestic sugar supplies and the consequent softening of its retail prices, futures trading was resumed on December 27, 2010. Second, the government imposed stocking limits for six months in August 2009 to rein in prices. In February 2010, the stock limits were extended for another four months. These limits barred bulk consumers from stocking sugar equivalent to more than 10 days of their consumption needs.

In terms of trade policies, the government response to rising prices has been to ban exports and liberalize imports. The export of sugar was banned in September 2008, while duty-free imports of raw and white sugar were allowed respectively in February and April 2009. The government further exempted imported sugar, both raw and white from the levy obligations and

FIGURE 20B. Sugar: Monthly Inflation Rate

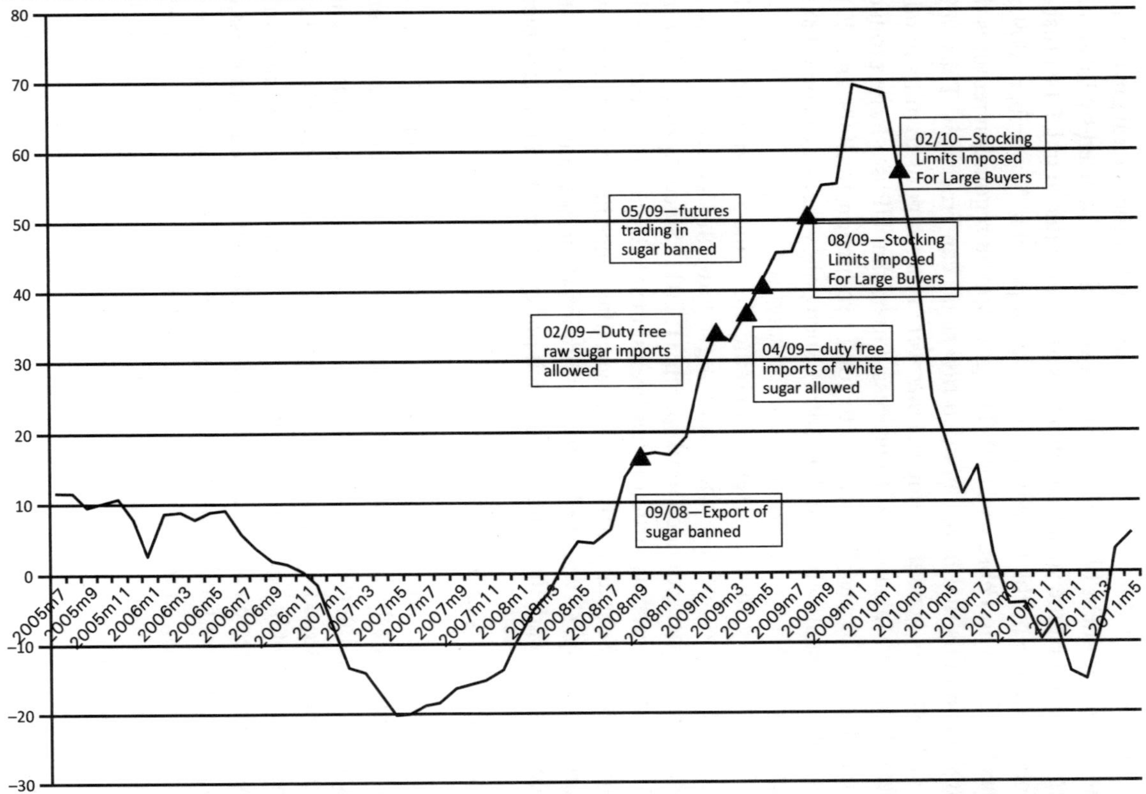

Source: Ministry of Commerce and Industry, Government of India.

quota restrictions on releases to the market. Given that inflationary pressures in sugar started emerging as far back as May 2007 (Figure 20B), the responses on the trade side could be considered to be delayed.

Importantly, even when imports were allowed by the central government, the UP government blocked imported raw sugar to be used by local mills arguing that first the sugarcane that was locally produced in UP be crushed. This restriction by the UP government was imposed in November 2009 and was lifted only in February 2010. The government regulation requires that only those mills that have imported raw sugar can process it. Thus, mills in UP could not send their imports that were lying at ports to mills in other states (e.g., in Maharashtra) to process. Given that UP is a large producer (contributing 26 and 27 percent to total production in 2009 and 2010, respectively, next only to Maharashtra), such policy measures could have a significant effect on market prices in the short run.

More recently, through a series of notifications, the government extended duty-free imports of raw and white sugar through December 31, 2010, which was further extended until March 31, 2011. Since April 1, 2011, the 60 percent import duty structure has been reinstated. On the export side, the ban stayed till February 2010 where it was lifted once inflationary pressures started receding. Given the history of high inflation in sugar as well as the recent rising trend, the rationale for such policy reversals remains unclear.

Cereals—Wheat and Rice

Although cereals are not the top direct contributor to overall food inflation, they are particularly important to study for various reasons. First, the government has been intervening extensively (along the supply chain in pricing, procurement, stocking, transport, and distribution) in the cereals markets. The government has also been the most proactive in wielding off any inflationary pressures on cereals. Second, the cereal price rises can put pressure on prices of other food items, through, e.g., substitution in production away from noncereals. Third, the rural wages could be linked to prices of cereals; hence, a rise in its price could result in a rise in the cost of production for other food items and for nonfood products as well.

The evolution of inflation rate for rice and wheat, and its potential correlates is shown in Figure 21A. Four key facts emerge from the data: (*a*) In general, except for the two peaks (1999 for both rice and wheat; 2006 for wheat and 2008 for rice), inflation in cereals has been largely muted; (*b*) during the peak of the global food price crisis, the cereal prices in India continued to be low; (*c*) the correlation of domestic rice and wheat price

FIGURE 21A. Cereals

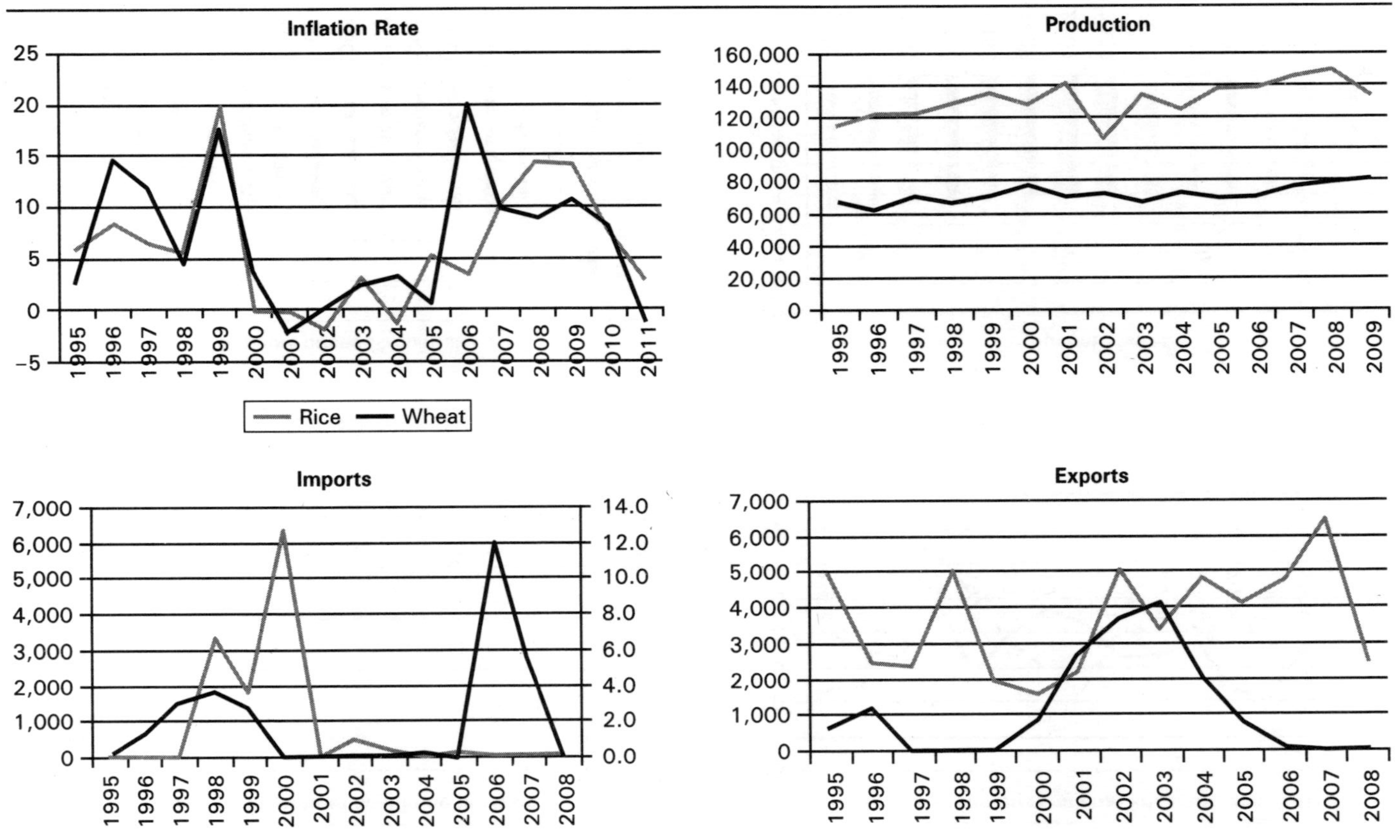

(Figure 21A continued)

(*Figure 21A continued*)

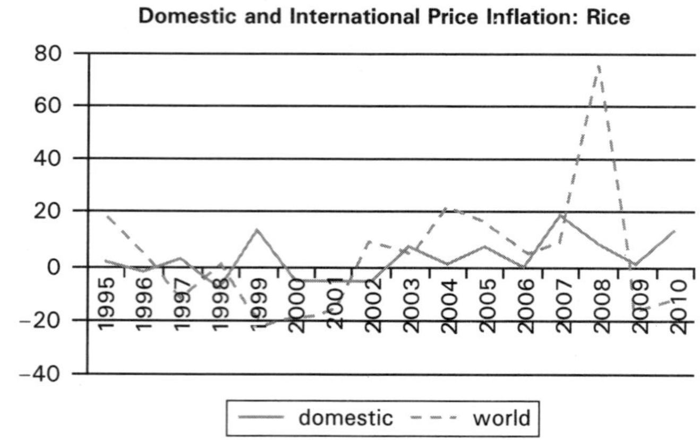

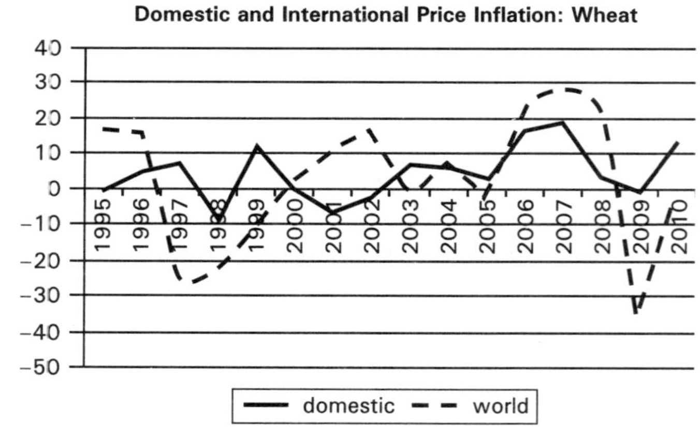

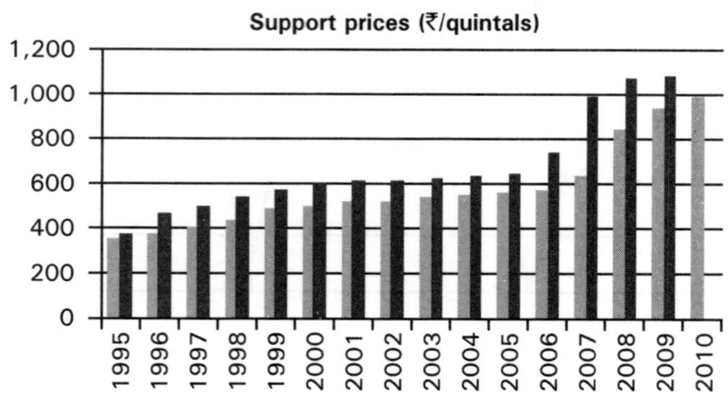

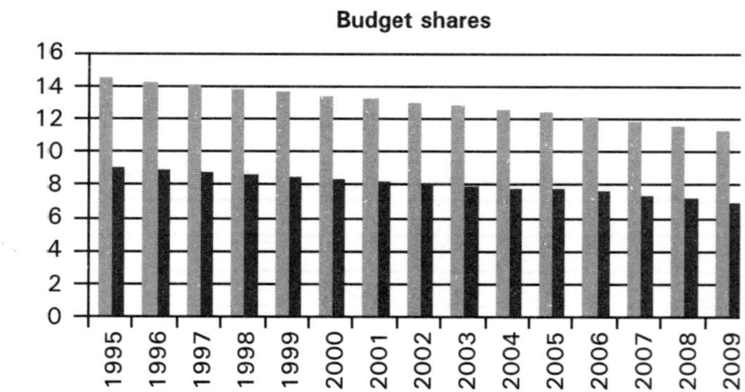

Source: Ministry of Commerce and Industry, FAOstat, Indiastat and IMF. The support prices are for fiscal years.

inflation with changes in international prices is weak (0.14 and 0.31 for rice and wheat, respectively); (*d*) since 2009, the inflation rate in both rice and wheat have been declining, unlike the high value items.

Long-term Factors The production of cereals, in particular wheat, has seen only modest increases over time (Figure 21A). Technology adoption in cereals has been slow with improved varieties of seeds exhibiting very sluggish diffusion. The area allocated to rice declined by 3 percent in the last decade (2000–10). The growth rate of production subsided on average from 2 to 1.6 percent per annum between 1991–2000 and 2001–10, and there was a marginal increase in yield growth rate from 1.3 to 1.6 percent during the same period. The picture in case of wheat has been gloomier. The growth rate in production and yield between the two decades has been reduced by one half and one third, respectively. Overall, the significant growth rates in cereal production that were achieved during Green Revolution have now become stagnant in the last decade or so. This has been attributed among other reasons to excess use of fertilizer (particularly urea that is heavily subsidized) and dropping groundwater levels owing to suboptimal crop choices.

In fact, compared to wheat and rice, coarse cereals have performed much better with nearly 4 percent increase in yields in the last decade. Other lagging crops such as oilseeds and pulses have also outperformed cereals in terms of yields (only marginally in case of pulses). In fact, the yield growth across all principal crops is unmatched with cotton where it grew by as much as 11 percent in the last decade (attributed to the spread of Bt Cotton). Among the important crops that showed worse performance than main cereals are sugarcane and some disaggregated category of pulses.

On the domestic policy side, government has intervened heavily in the cereal markets through three main policies: (*a*) procurement, (*b*) stocking, and (*c*) releases through the PDS.

First, the procurement system in the form of MSPs that has the objective of ensuring reasonable income to farmers and adequate availability of foodgrains to consumers at reasonable prices plays an important role in determining food inflation. MSP acts as a floor price and serves as a benchmark for inflation. Importantly for MSP to be effective in procurement, it has to rise above the market clearing price else the government will not find any sellers. Hence any increases in MSP can set up inflationary pressures in the system. Gaiha and Kulkarni (2005) showed a strong positive correlation between the level of MSP (for rice and wheat) and WPI as well as CPI-AL after controlling for time trends and level of income.

Though MSP is aimed at providing incentives to farmers, it traditionally has been below international prices. This implies an implicit tax on the farmers, given the regulations on exports. Further, as the fiscal costs of the MSP system have spiraled, a long-term effect has been the declining public investment in agriculture (Gaiha and Kulkarni, 2005).

Second, the government maintains both buffer stocks and strategic grain reserves (the latter since 2008). The stockpiling and the release of reserves are tools used by the government for price stabilization. Third, on the side of disbursement, by keeping the variation in issue prices to the PDS low, government has tried to stabilize prices for the consumers. The PDS, however, is mired with leakages and corruption (Kotwal et al., 2011) and does not serve as an adequate tool for shielding consumers from price pressures.

On the trade side, there was a ban on exports of both wheat and rice till the mid-1990s. For most of the period we focus on in this paper, the restrictions on exports were already lifted. However, similar to milk and sugar, the tariffs on both rice and wheat are nearly prohibitive at 70 and 50 percent, respectively.

With significant public intervention, the 1990s witnessed some small steps at reforming the food policies of the government. The government has more or less removed the restrictions on interstate movement of commodities and proposed to do away with the Essential Commodities Act and replace it with an emergency act. Since, agriculture is a state subject, the government has advised states to amend laws such as Agriculture Produce Marketing Committee act (APMC). In reality most of the restrictive policies have remain unchanged (see Jha et al. 2010).

Short-term Factors One of the short-term factors contributing to inflation in cereals has been weather shocks, e.g., a drought in 2009 in North India. As shown in Figure 21A, the drought affected mainly the production of rice, which declined by 10 percent.

While the existence of a support price system constitutes a long-term factor for explaining inflation as discussed earlier, we consider revisions in MSP as a short-term cause of inflation. In principle MSP is based on cost of cultivation (C2) calculated by CACP that takes into account all expenses in cash and in kind, plus rent paid for leased land, imputed value of family labor and interest costs on working and fixed capital. Since 1997–98, the MSPs have often been set higher than the C2 benchmark (for political reasons). In 2001–02, e.g., the weighted average C2 costs of eight wheat-producing states was ₹4.83 per kg while the MSP was set at ₹6.20 per kg (Gaiha and Kulkarni, 2005).

MSP generally gets revised upward every year, but the magnitude of increase can vary and sometimes substantially. For example, the MSP for both wheat and rice increased by one-third in 2007–08 and 2008–09, respectively. The increases in MSP have moreover tended to be higher in high-inflation years in order to maintain procurement. For example, in case of wheat, the sharp increases in MSP by 15 and 33 percent in 2006–07 and 2007–08 coincided with the steady rise in inflation rate throughout 2006, and the first half of 2008 (Figure 21B). Ideally MSP should be low to reduce procurement in bad (high inflation) years and should be high in good (low inflation) years to procure more and stock up as argued by Basu (2010). The absence of any downward revision in MSP over time and the lack of adjustment for inflationary pressures make it an important factor accounting for inflation in cereals.

Another domestic policy that the government has used has been the restriction on futures trading. For example, in February 2007, in order to reduce incentives for hoarding, the government banned futures trading in rice and wheat (Figure 21B).

Further, one of the main short-term measures to address inflationary pressures in cereals is trade policy, i.e., allowing imports and restricting exports during times of adverse production shock. The peak in wheat inflation in November 2006 of nearly 30 percent was followed by immediate action by the government where it imported nearly 6 million tons of wheat, and it did put significant downward pressure on prices. Similarly, concomitant with the May 2008 peak in the inflation rate of nearly 14 percent, the government again imported 1.7 million tons in the month of May itself. Likewise, export bans were placed on wheat in February 2007, immediately after the inflation rate peaked; and on non-basmati rice in April 2008 in the midst of high and rising inflation. Overall, the trade policy response in cereals is distinctive in terms of its promptness in reaction to rising inflation and in stark contrast to the case of sugar and milk.

Edible Oils

The edible oil sector in India comprises vanaspati, mustard and rapeseed oil, soybean oil, palm oil, sunflower oil, safflower oil, and others. According to the FAO stat, India has consistently been ranked among the top four producers of oilseeds and top consumers of edible oils in the world. It is now the biggest importer of edible oils overtaking China in 2009. Specifically the imports include soybean oil mainly from Brazil and Argentina and palm oil from Indonesia and Malaysia. Over the long run, imported oils mostly palm

FIGURE 21B. Cereals: Rice and Wheat: Monthly Inflation Rate

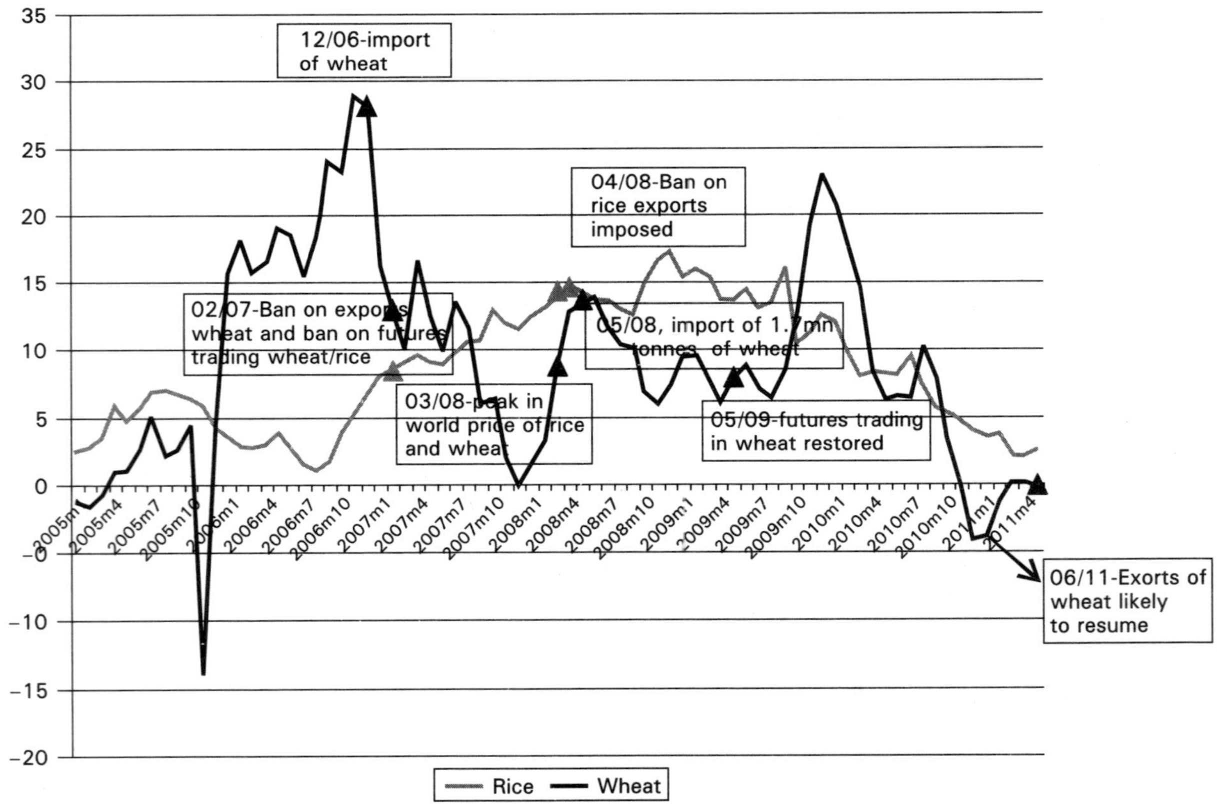

Source: Ministry of Commerce and Industry: Government of India.

and soybean have accounted for most of the consistent growth in consumption and now account for about half of the oil consumed in India.[29]

Among the edible oils, we focus more on vanaspati and mustard and rapeseed oil, given that they appear frequently as top contributors to the overall food price inflation (Figure 18), both are among the top five in 2003, and are in the top ten in many years. These two edible oils, however, are different from others in terms of high domestic production and comparatively low import penetration (particularly in the case of mustard and rapeseed oil).

The pattern of inflation in these two commodities (Figure 22) shows that in the last decade there have been three key peaks in their inflation rates. The first one occurred in 1998 for mustard and rapeseed oil when its inflation rate jumped to more than 40 percent; the second in 2002 and 2003, when the inflation rate for both vanaspati and mustard and rapeseed oil reached close to 20 and 40 percent, respectively. The third peak for both which was relatively muted corresponded to the global food price crisis in 2008. Other than these peaks, the inflation has been quite low. Recently, since September 2010, there has been some upward pressure on prices of both vanaspati, and mustard and rapeseed oil with their inflation rates in May reaching 10–11 percent.

Long-term Factors The major long-term factors responsible for the high consumption and import of edible oil in India are increases in per capita income and population, change in tastes and preferences, as well as low productivity in the domestic oilseed sector.

In terms of policies, imports were liberalized in 1994 with edible oils taken off the negative list and put under Open General License. Since then, an increasing portion of the edible oil consumption started getting sourced from imports (Dohlman et al., 2003). The per capita availability of edible oils went up drastically after liberalization. In the decade prior to liberalization, the growth of about 6 percent occurred in per capita availability while in the decade after, it went up by more than 60 percent.

The liberalization in edible oils occurred in phases. Based on Lahiri Committee Report (2006) on edible oils, two distinct phases are discernible: the first between 1994 and 1998, when customs duty on edible oils progressively came down to reach a low of 15 percent in July 1998; and the second after 1999, when such duties witnessed an upward trend, touching a high of 92 percent for refined palm oil in April 2005.[30] Between 1994 and

29. Source: Directorate of Vanaspati, Vegetable Oils and Fats, Government of India.

30. We thank Ashok Lahiri for pointing this out.

FIGURE 22. Edible Oils

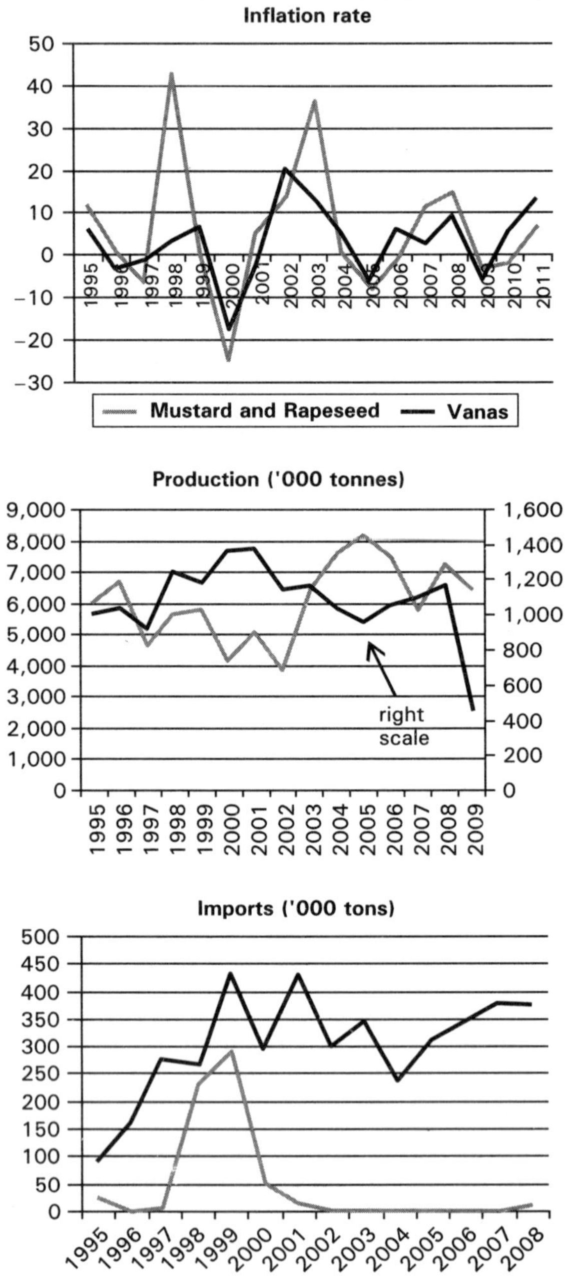

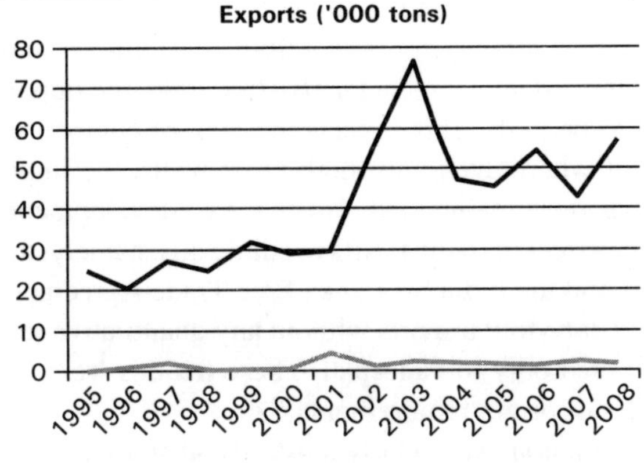

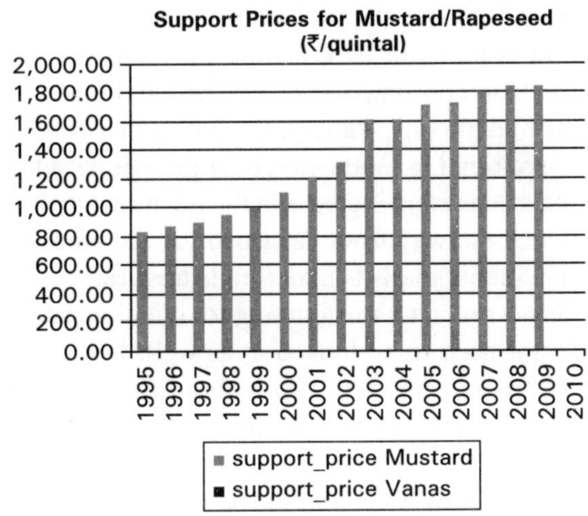

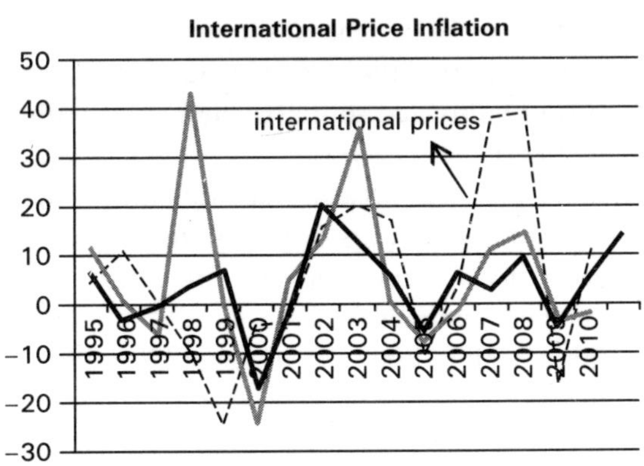

Source: Ministry of Commerce and Industry, Government of India, FAOstat, Indiastat and IMF. The international prices are average for all edible oils.

2005, import duties were revised 17 times (Table A2 from Lahiri Committee Report), creating uncertainty for both edible oil producers and importers.

With these upward revisions in tariffs after 2000, the import penetration has continued to be high. This reflects the lack of competitiveness of the edible oil sector where even with high tariff walls, the import's share in consumption has been maintained. Further, two developments could have counteracted the effect of rising tariffs in edible oils, namely, the Indo-Nepal Treaty on trade and the India Sri Lanka Free Trade Agreement. Both these treaties allowed duty-free imports without any quantitative restrictions; the latter was restricted only to vanaspati. These treaties were further revised to relax the rules of origin requirements.[31]

Edible oil production, particularly mustard and rapeseed oil, has typically been fluctuating in line with the variability in oilseed production, which partly explains fluctuations in its inflation rate. This is possibly because the share of imports in consumption of mustard and rapeseed oil is much lower than in other edible oils. Vanaspati, on the other hand, has exhibited a distinct downward trend in production over the last decade. Yet, this has not translated into a consistent increase in the inflation rate for two reasons: (*a*) shifting preferences away from vanaspati toward other edible oils and (*b*) imports from regional markets under trade treaties.

Among the long-term factors accounting for inflation patterns, outcomes in the oilseed sector need to be looked at as well. India's trade policy in oilseeds has been quite restrictive with high tariffs (of about 30 percent) and nontariff barriers such as the need for proof of genetically modified (GM) free status.

On the domestic side, over the long run even though India has been one of the largest producers of oilseeds in the world, the productivity in the sector has been quite low (50 percent of the world average). More importantly, the yields for domestic oilseed production have been quite variable as oilseeds are mostly cultivated in unirrigated areas with low quality of seeds (Srinivasan, 2005).

The increase in the prices of oilseeds has been higher than that in edible oils, partly as a result of deliberate policy of giving higher support prices to oilseeds to encourage diversification away from cereals. However, oilseed cultivation has remained unattractive and since 2000, the area under oilseeds has stagnated (after a peak in 2004–05 of about 27 million hectares, it has hovered around there).

31. On March 2, 2002, the Treaty was modified by incorporating a yearly import limit of 1 lakh tons to address the concerns of Indian vanaspati industry.

Short-term Factors Historically, the two spikes in the inflation rate for mustard and rapeseed oil in 1998 and 2003 coincide with production shocks, but importantly also with significant increases in the support prices for oilseeds. The support prices for mustard and rapeseed went up by over 20 percent between 2002–03 and 2003–04. This coincided with the sharp rise in world prices around the same time.

Another short-term factor in explaining the patterns of inflation is the movement in international prices. The domestic prices of mustard and rapeseed oil and vanaspati like all other edible oils track the international prices quite closely. The international prices of edible oils have shot up recently, with an average y-o-y inflation of 32 percent between September 2010 and May 2011. Although total imports of edible oils by private traders also increased quite significantly during this period, this did not result in lower inflation rates in general due to the high prices at which the imports came in.

Fruits and Vegetables

Fruits and vegetables typically exhibit significant variability in production and yields over time. Weather shocks in the principal production centers tend to have significant effects on the prices of vegetables in particular. Among vegetables, the highest spikes have been observed in the case of onions followed by potatoes, while among fruits, rise in prices of mangoes have been most pronounced (Figure 23). In some years the price rises in onions and potatoes have tended to coincide (e.g., in 1998), possibly because they are grown in the same regions (Karnataka and Maharashtra), and also during the same seasons. One of the striking features of inflation in the fruits and vegetables category is that the magnitude of the peaks in inflation at times for some products have gone into triple digits.

Inflation in fruits and vegetables exhibits two distinctive features. First, looking at monthly inflation patterns, the seasonality in prices of fruits and vegetables is quite evident. All fruits and vegetables tend to have comparatively high prices in some seasons compared to others. Second, there are wide swings in prices, implying that there are both periods of scarcity as well as glut in the market within short time frames. This happens probably because supply tends to be seasonal, while demand does not (e.g., in case of onions).

Long-term Factors Both demand- and supply-side factors could account for the inflationary pressures in fruits and vegetables. As discussed earlier, there

FIGURE 23. Fruits and Vegetables

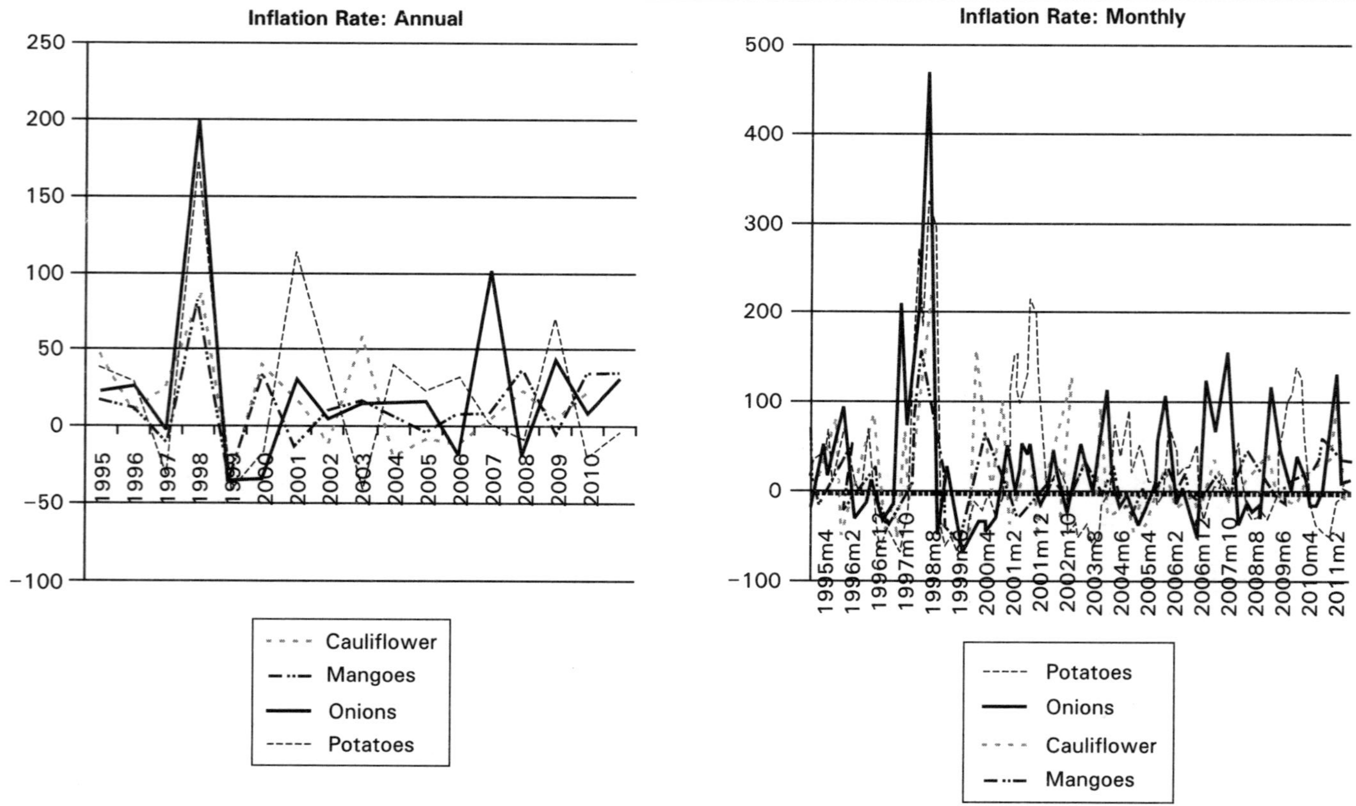

Production of Onions ('000 tonnes)

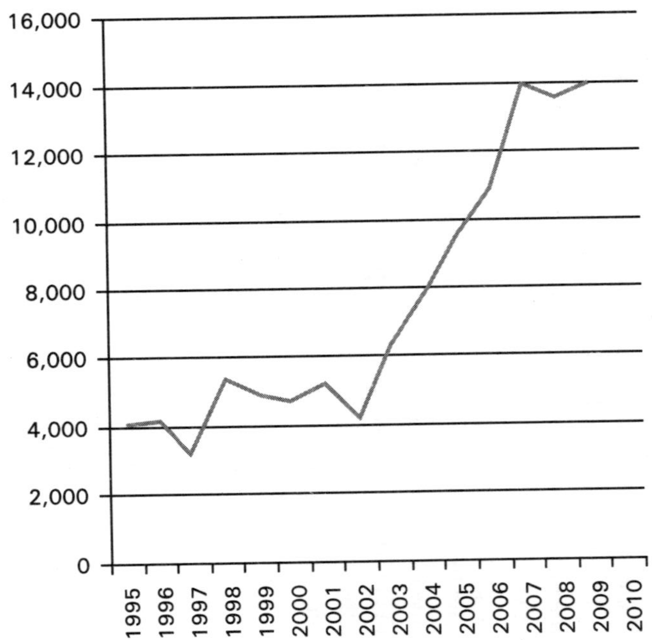

Trade in Onions ('000 tonnes)

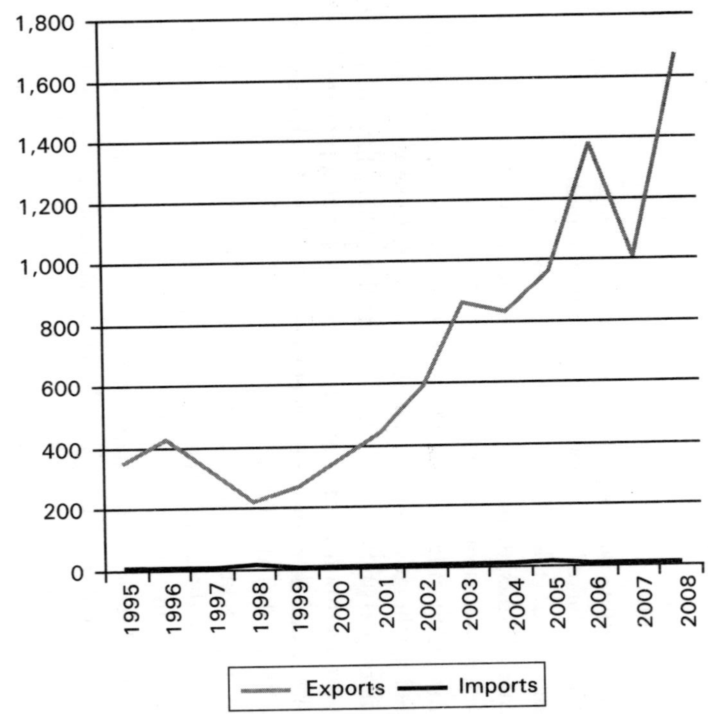

Sources: Ministry of Commerce and Industry, Government of India, FAOstat and Indiastat.

has been a continuous shift in demand patterns toward high-value products that include fruits and vegetables. Some of the increases in demand for vegetables are linked to the demand for other commodities, e.g., the demand for onions in part is derived from increases in meat consumption. Similarly, in the case of fruits, there could be links to the rising demand for processed food such as juices and ice creams.

On the supply side, the usual factors where noncereal sectors have been neglected by the government can be considered to be responsible for the inflationary pressures in fruits and vegetables. There has been a gradual shift in production toward fruits and vegetables, but it probably is slower than what would be required to meet the needs from the changing demand patterns (Birthal et al., 2007).

The important constraints for fruits and vegetables are most likely off-farm where there has been very slow progress. There is an inadequate capacity for cold storage and there are no cold chains or drying technologies. Seasonal shocks in the main production centers thus increase the prices instantly. In addition, the supply chains remain uncoordinated with a host of middlemen that add to markups throughout the chain. Also, the benefits of the increases in prices in most part do not accrue to the farmers. The lack of coordinated chains also implies that there is high incidence of waste and spoilage, the importance of which gets magnified in times of scarcity. The long-term policies on retail and processing have played a role in creating this situation. The wide swings in the prices of fruits and vegetables also suggest that the market information systems have historically not been well developed in the country.

Onions is a particularly interesting and important case to study, given the frequent and high peaks in its inflation rate. Moreover, it has attracted considerable media and policy attention. Figure 23 shows that both production and productivity of onions has gone up significantly in recent years. Yet, owing to the declining agricultural research and extension systems, the productivity in India remains extremely low in international comparison. According to the FAO, compared to 51 tons per hectare in the US and 22 tons per hectare in China, India's productivity was only 13 tons per hectare in 2006. Nevertheless, the production is usually sufficient to meet the domestic needs and the surplus is exported.

On the marketing side National Agricultural Cooperative Marketing Federation of India (NAFED) intervenes whenever there is a glut in the market and provides price support. Thus, though there is no consistent system of MSP for onions, NAFED does cover the downside risk to some extent. NAFED is also the agency that handles the export of onions and it fixes a minimum export price. Since exports are canalized (they have to necessarily go through NAFED), this could affect the incentives of producers to grow onions as an exportable crop.

SHORT-TERM FACTORS Because of inadequate transport, storage and processing infrastructure, seasonal variations lead to changes in prices of fruits and vegetables including in the case of onions. In 2010, in the main production center of onions in the winter season, i.e., Maharashtra, higher than expected rains destroyed the crops and there was an immediate push toward rising prices of onions starting end of the year. These rain shocks could also affect seeds and saplings, and thus affect crops for the immediate next period and give short term persistence to inflationary pressures. Since onion is an exportable crop, the government responded by imposing a ban on exports of onions in December 2010. At the same time, the government allowed the imports of onions from Pakistan. Moreover, government raided many warehouses of traders to prevent hoarding. In an earlier inflation episode, toward end of 1998, the government had also banned exports of onions to stem price increases.

The rapidity in response of the government toward checking the rise in prices is exceptional for a noncereal foods category. Possibly the events in 1998 where there were strong political fallouts from rise in onion prices pressured the government to act quickly.

Conclusions and Policy Implications

This paper uses a disaggregated commodity-level and high-frequency dataset to provide a forensic account of inflation in India over the last two decades, with a focus on food prices. This forensic analysis comprises several elements.

We start by discussing the measurement of inflation in India, the different indices officially published, and outline the measurement of food price inflation. We then discuss the trends in overall inflation. We establish the commonly overlooked fact that low inflation has been a rare occurrence in the Indian economy in the last two decades. Importantly, the long-term trend in the inflation rate exhibits a U-shaped pattern with a structural break in 2000 and an inflection point in 2002 (2003 for CPI).

Next, we show that the long-term trend for food inflation has also followed a U-shaped path over the last two decades, with a clear reversal of the declining trend starting early 2000s. In addition, there is evidence for a moderate correlation between international and domestic food price inflation, albeit with significant variation across commodities based on tradability, e.g., edible oils exhibit a higher degree of pass-through. Moreover, the correlation is lower when world prices are high than when they are low.

Further, we distinguish between food and nonfood inflation based on WPI and find that food price inflation is typically higher than nonfood (with the long-term trend for food always above nonfood). Moreover, its persistence is close to that of nonfood. There is also evidence for a long-run cointegrating relationship between food, nonfood and overall inflation rates; and based on both a simple linear regression as well as a framework incorporating the dynamic behavior of inflation, we find that there is a significant pass-through from food to nonfood inflation rates. A one-unit shock to food inflation which translates to a 0.2 percentage point increase in the long run is associated with at least a 0.05 percentage point increase in nonfood inflation.

Finally, we also establish that the relative price of food and aggregate inflation co-move with each other post 2000, although this finding does not survive in simple regressions (with CPI-based inflation measures, we get a more robust relationship).

Given the importance of food in overall inflation, we take a step forward and rather than treating food as a monolith, we disaggregate it into its different components. We explicitly quantify the contribution of specific commodities to food price inflation. Although there is some variation over time, the top contributors are typically the following: milk and fish, sugar, rice and wheat, onions, potatoes, and to a lesser extent, edible oils. Moreover, the increase in relative price of food since mid-2000s has been driven primarily by high value food items—likely to be explained by the changing demand patterns. In contrast, the increase in the 1990s was influenced primarily by the economic reforms that improved the terms of trade for agriculture.

Finally, we conduct case studies of specific items from the list of top contributors to study the inflation patterns and the factors explaining them. The choice of commodities is motivated by each one of them being been among the top contributors to food price of inflation at some point during the last two decades. Importantly, the experiences of each of these commodities tell a distinct story about the factors explaining the inflation patterns and the policy responses. For example, persistent demand-supply gap in milk owing to changing tastes of consumers; post-Green Revolution fatigue reflected in declining or stagnant productivity in cereals; multipronged government interventions in sugar; high import penetration in edible oils; and inadequate infrastructure for smoothing prices (across space and over time) for fruits and vegetables.

Some unifying messages also emerge from the case studies. We find that long-term structural factors and lasting government policies as well as short-term exogenous shocks and temporary government responses could explain the patterns of inflation across commodities. Given the built-in inflationary pressures due to the existence of long-term factors, short-term shocks can create a magnified effect on inflation. In addition, the intensity and speed of government responses determines how quickly and to what extent the inflationary pressures are reined in. Moreover, commodity-specific policies have both direct effects on the targeted sectors as well as possible spillovers to other sectors (e.g., the cereal-centric policies of the government have effects on supply of high-value items).

The findings in this paper can have several policy implications. First, combining both the macro and the micro analysis, a case can be made that in the context of a developing country like India, both the central bank and other branches of the government can play a meaningful role in tackling the inflation problem.[32] Below we discuss some specific policy recommendations.

Given the high weight of food in overall inflation and the importance of supply factors or more structural and less interest-sensitive demand factors in explaining food price inflation, the hope is that the RBI can control overall inflation by curbing the demand for nonfood items that are more interest sensitive.[33] However, even if the monetary transmission mechanisms operate efficiently, controlling nonfood inflation may have a limited effect on overall inflation, particularly if the central bank envisages to target an inflation rate with CPI-based indices which have a relatively small weight on nonfood (31–54 percent).

Therefore, one pertinent question is can monetary policy play any role in alleviating food price inflation? There are several possible mechanisms. First, if monetary transmission through the exchange rate channel is effective; i.e., if tighter monetary policy results in an appreciated exchange rate, the prices of imported food items in local currency would fall. The strength of this mechanism would vary across commodities depending on the degree of import penetration (which itself could be endogenous to inflationary pressures as shown earlier). For example, we can expect this channel to be operative to a greater degree in edible oils. The overall importance of this

32. See Jha (2005) for a similar argument.

33. Kapoor and Ravi (2009) show using micro data for India that the demand for food items is insensitive to interest rate changes.

channel would therefore depend on the pass-through to exchange rate as well as the importance of highly tradable commodities in the food price index.

Second, tighter monetary policy would reduce the aggregate demand and incomes of households and, in turn, reduce the demand for food. The strength of this mechanism would depend on monetary transmission, which itself is likely to be "weak and unreliable" in developing countries (see Mishra et al., forthcoming, 2012; Mishra and Montiel, forthcoming, 2012). In addition, the strength of this channel would depend on the income elasticity of demand for food, which existing studies estimate to be less than one (e.g., Gandhi and Zhou, 2010; Mittal, 2006). Therefore, if transmission is weak, and/or if food demand is relatively income inelastic, monetary policy is likely to have only limited effects in curbing food price inflation through this mechanism.

Alternatively, one can think of another commonly ignored channel which could have an opposite effect, i.e., where monetary tightening leads to a rise, rather than a fall in food prices in the short run (the so-called "price puzzle"). One of the explanations for the price puzzle considered in the literature is the existence of a "working capital channel" wherein a rise in interest expenses increases the marginal cost of production (Christiano et al., 1997; Barth and Ramey, 2001). This channel could potentially work all along the value chain from farm to retail (including processing, transportation, etc.), and is likely to be stronger in the case of high-value food items such as livestock products (von Braun et al., 2005). The availability of working capital has also been identified as among the five major constraints affecting agro-processing industry in India (Kachru, 2006). This channel has to our knowledge been largely ignored in the academic and policy discussions. If the working capital channel is quantitatively important, then this would call for a less aggressive monetary policy stance.

The relative strength of these three mechanisms is an empirical question and merits further research. To the extent that the food demand is less interest rate sensitive; the exchange rate and aggregate demand channels are relatively weak; and the working capital channel is important, it is unlikely that tightening monetary policy can relieve food price inflation.[34]

Aside, the paper also has implications for which price index the central bank should target—headline or core. Given that food has a high weight in

34. A reason other than the working capital channel which could make the case for monetary tightening weaker is when higher interest rates attract capital inflows and lead to overheating, and high overall inflation. For example, Turkey in the recent past reduced interest rates in times of rising inflation in order to stem capital inflows.

the consumption basket and that food price inflation has been consistently above nonfood by itself calls for focusing on headline rather than core measures of inflation. This is supported by further evidence of persistence in food price inflation, and the existence of some pass-through from food to nonfood. This policy message is consistent with previous papers, e.g., Walsh (2011), Anand and Prasad (2010), and Catao and Chang (2010), and is also in line with the RBI's view.[35]

Furthermore, regarding which measure of headline should the central bank use—given that using the WPI significantly underestimates the degree of association between relative price of food and aggregate inflation—points to the fact that as an index relevant to consumers, CPI is more appropriate. Thus, a strong case for developing a more comprehensive CPI can be made.

Some important policy lessons also emerge for other government branches from the commodity-level analysis. Trade barriers are typically very high and often prohibitive for most food items, including for the top drivers of food price inflation. Yet, in the case of almost all commodities we studied, trade policy (e.g., export ban, duty free imports, etc.) has been used as a shock absorber to cool inflationary pressures. One exception is edible oil which was liberalized in the mid-1990s, and where inflation has been quite muted. This raises the question whether the patterns of inflation we observe in most commodities would have been different if trade were more liberalized to begin with.

The case studies further suggest that the effectiveness of trade policy as a stop-gap tool to curb inflation depends on the timing of the response, e.g., political expediency resulted in a rapid response in the case of onions and cereals, while in case of milk and sugar, it was much delayed. Promptness in the implementation of trade policy measures is crucial for two reasons. First, it takes time to build new trade relationships if none existed from before (e.g., for clearance to import livestock products, a 30-day advance notice is required). Second, the international prices, particularly for large buyers like India could change affecting the import prices.

Based on the case studies, one clear policy message to emphasize is that curbing inflationary pressures could be a rationale in itself to reduce long-lasting import barriers in agriculture (both tariff and nontariff barriers).[36]

35. See for example, Subbarao (2011).

36. Interestingly, India for a long time had negotiated zero bound tariffs for many agricultural commodities, e.g., rice and dairy products under the Geneva Protocol (1947). The bound tariffs were later renegotiated (as late as 2000) after the Uruguay Round (e.g., to 80 percent and 60 percent in rice and milk, respectively).

Temporary and targeted export regulations could still be warranted from the perspective of controlling inflation (although long-term application of export regulations need to be avoided as it could generate perverse incentives against raising production and productivity).

The overarching message that emerges from this paper is that multiplicity of instruments (e.g., different forms of support prices, domestic subsidies, futures trading) as well as the number of government agencies involved makes the task of controlling inflation a daunting challenge. The in-depth case studies suggest that, in general, policies and institutions primarily created for increasing availability and stabilizing prices—such as relatively closed borders and domestic restrictions—could have actually aggravated the food inflation problem. Therefore, the recent episode of food price inflation could reopen the debate over whether a more open and less interventionist stance of the government can stabilize food prices and, hence, overall inflation. Finally, the complex web of multiple agencies and multiple instruments working potentially at cross-purposes can make the task of enhancing the effectiveness of monetary policy an uphill one.

Overall, we recognize that this paper barely scratches the surface for many important topics related to the issue of inflation in India—both at the macro as well as at the micro level. The goal of this paper is to set the stage for research on many important issues raised above. In particular, further research is warranted to study the importance of the working capital channel. Further, quantifying the effects of different factors/policies in explaining the patterns of inflation across commodities and over time is also an important area for further research.

APPENDIX

TABLE A1. Classification of Commodities By Inflation and Weight

	High inflation		Low inflation	
	High weight	*Low weight*	*High weight*	*Low weight*
1998	**Onions (186.2,0.8)**	Bagasse (345.6,0.2)	Rice (5.6,8.3)	Bread and Buns (5.2,0.3)
	Potatoes (152.8,0.9)	Garlic (60.3,0.3)	Wheat (4.4,5.2)	Cakes and Muffins (5,0.1)
	Mangoes (75.2,3)	Turmeric (41.7,0.4)	Cattle Feed Gola (4.2,0.5)	Ragi (4.4,0.1)
	Cauliflower (73.1,1.1)	Black Pepper (35.3,0.1)	Poultry Chicken (4.1,1.9)	Cardamoms (4.3,0.1)
	Cabbage (50.7,0.9)	Grapes (28.9,0.4)	Mutton (3.9,1.6)	Urad (4,0.5)
	Oranges (45.3,0.6)	Sapota (26.9,0.2)	Sugar (3.3,8.1)	Ginger (3.8,0.2)
	Mustard and Rapeseed Oil (42.3,2.1)	Papaya (26.7,0.5)	Blended Tea Leaf (3.1,1.3)	Dry Ginger (3.5,0.2)
	Tomatoes (38.8,1.2)	Gingelly Oil (18.4,0.2)	Vanaspati (3.1,3.3)	Cummin (2.6,0.2)
	Rice Bran Oil (34.9,0.9)	Tea (18.4,0.5)	Maida (2.6,2.1)	Jowar (2.2,0.4)
	Soyabean Oil (34.9,1.8)	Mustard (15.2,0.2)	Gram (1.9,1.6)	Rawa Sooji (1.4,0.4)
	Dry Chillies (31.8,0.7)	Gur (13.3,0.4)	Atta Wheat Flour (0.9,1.8)	Coffee Powder (0.2,0.1)
	Arhar (29.5,0.6)	Khandsari (12.7,0.3)	Bajra (–0.4,0.5)	Canned Fish (0,0.3)
	Brinjal (25.7,1.4)	Apples (11.9,0.5)	Beef and Buffalo Meat (–1.1,0.5)	Sweet Potatoes (–0.5,0.1)
	Unblended Tea Leaf (23.3,0.8)	Masur (11.6,0.3)	Powder Milk (–1.7,0.9)	Wheat Bran (–1.1,0.4)
	Okra (23.1,0.6)	Moong (11.1,0.4)	Egg (–2.5,0.9)	Coffee (–1.7,0.3)
	Cotton Seed Oil (20.2,1.2)	Salt (10.5,0.2)	Banana (–7.3,1.6)	Pork (–2.2,0.3)
	Groundnut Oil (19.8,1.4)	Butter (9.4,0.3)	Maize (–7.9,1)	Betelnuts, Arecanut (–5.8,0.5)
	Sunflower Oil (16.1,0.8)	Pineapple (6.2,0.2)	Coconut (–20.3,1.1)	Barley (–8.8,0.1)
	Inland Fish (10.5,2.7)			Copra Oil (–10.1,0.5)
	Ghee (9.6,1)			Tapioca (–12.3,0.3)
	Milk (8.5,15.1)			Groundnut (–16.4,0.2)
	Marine Fish (8,3.4)			Green Peas (–21.4,0.5)
	Biscuits (7.9,1.6)			Guava (–27.4,0.4)
	Cashwenuts (7.8,0.8)			Corriander (–31.9,0.1)

(*Table A1 continued*)

(*Table A1 continued*)

	High inflation		*Low inflation*	
	High weight	*Low weight*	*High weight*	*Low weight*
2008	Tea (38.5,0.5)	Corriander (84.2,0.1)	**Sugar (6.5,7.1)**	Maize Atta (6.3,0.1)
	Mangoes (37.4,2.7)	Ginger (76.8,0.2)	Green Peas (5,0.5)	Khandsari (6.3,0.3)
	Others (30.7,0.8)	Groundnut (45.3,0.2)	Cashwenuts (4.7,0.7)	Fish Meal (6.2,0.2)
	Papaya (25.8,0.4)	Masur (29,0.2)	Cotton Seed (4.6,0.5)	Black Pepper (6.2,0.1)
	Cauliflower (24,1)	Mustard (28.5,0.2)	Food powder (4.5,0.6)	Jowar (6,0.4)
	Tomatoes (23.5,1.1)	Excluding Oil (26,0.2)	Bajra (4.5,0.5)	Copra Oil (5.4,0.4)
	Brinjal (17.5,1.2)	Cardamoms (25.9,0.1)	Biscuits (4.3,1.4)	Besan Gram Powder (4.7,0.4)
	Cotton Seed Oil (15.8,1.1)	Gingelly Oil (24.5,0.2)	Maize (4.2,0.9)	Ragi (3.9,0.1)
	Mustard and Rapeseed Oil (14.8,1.9)	Coffee Powder (23.7,0.1)	Atta Wheat Flour (4.1,1.6)	Condensed Milk (3.3,0.2)
	Rice (14.2,7.4)	Salt (22.4,0.2)	Blended Tea Dust (4.1,0.6)	Ice Cream (3.3,0.2)
	Arhar (13.9,0.6)	Dry Ginger (19,0.2)	Powder Milk (4.1,0.8)	Rawa Sooji (3.3,0.3)
	Blended Tea Leaf (13.5,1.1)	Gur (18.6,0.3)	Dry Chillies (3.7,0.7)	Papad (3.2,0.3)
	Cattle Feed Gola (13.4,0.5)	Turmeric (18.2,0.3)	Maida (3.4,1.9)	Butter (3.2,0.3)
	Coconut (12.9,1)	Barley (18,0.1)	Egg (3.2,0.8)	Pork (2.7,0.2)
	Soyabean Oil (12.5,1.6)	Guava (17.6,0.3)	Groundnut Oil (2.8,1.3)	Bread and Buns (2.7,0.3)
	Mutton (12.1,1.4)	Unblended Tea Dust (16.8,0.4)	Cashew Kernel (1.5,1.6)	Roasted Cashew Nut (2.2,0.1)
	Oranges (11.7,0.5)	Others (16.6,0.2)	Beef and Buffalo Meat (0.5,0.5)	Lemon (1.1,0.3)
	Marine Fish (11.2,3)	Coffee (15.3,0.3)	Gram (0.5,1.4)	Pickles (0.7,0.1)
	Rice Bran Oil (10.5,0.8)	Rice Bran Extraction (15,0.4)	Inland Fish (–0.9,2.4)	Sugar Confectionary (0.4,0.2)
	Unblended Tea Leaf (10.4,0.7)	Tapioca (13.9,0.3)	Betelnuts, Arecanut (–1.2,0.4)	Wheat Bran (0.2,0.4)
	Rice Bran (10,0.5)	Bagasse (10.4,0.2)	Poultry Chicken (–2.3,1.7)	Vegetables Seeds (0,0.4)
	Molases (9.6,0.5)	Grapes (10.3,0.4)	Apples (–3.7,0.4)	Chire Beaten Rice Flakes (0,0.2)
	Vanaspati (9.4,2.9)	Canned Meat (9.9,0.1)	Cabbage (–4.3,0.8)	Canned Fish (–0.3,0.2)

	Banana (8.8,1.4)	Cakes and Muffins (8.8,0.1)	Processed Prawn (−5.1,0.5)	Sapota (−0.8,0.2)
	Wheat (8.8,4.6)	Pineapple (7.8,0.2)	**Potatoes (−16.9,0.8)**	Cummin (−1,0.2)
	Ghee (8.6,0.9)	Mixed Spices (7.4,0.3)	Onions (−21.4,0.7)	Litchi (−3.3,0.2)
	Palm Oil (7.8,1.7)			Moong (−4.2,0.3)
	Okra (6.9,0.5)			Sweet Potatoes (−6.3,0.1)
	Sunflower Oil (6.7,0.7)			Urad (−10.6,0.4)
	Milk (6.6,13.3)			Garlic (−25.2,0.3)
2009	Potatoes (70.9,0.8)	Turmeric (66,0.3)	Oranges (8.7,0.5)	Coffee Powder (8.7,0.1)
	Molases (53,0.5)	Gur (57.6,0.3)	Egg (8.3,0.8)	Rawa Sooji (8.3,0.3)
	Sugar (47.7,7.1)	Khandsari (44.8,0.3)	Processed Prawn (7,0.5)	Excluding Oil (7.3,0.2)
	Arhar (39.2,0.6)	Tapioca (42,0.3)	Beef and Buffalo Meat (7,0.5)	Jowar (7.2,0.4)
	Onions (36.9,0.7)	Moong (39,0.3)	Banana (6.5,1.4)	Maize Atta (7.1,0.1)
	Inland Fish (34.2,2.4)	Litchi (33.2,0.2)	Atta Wheat Flour (5.7,1.6)	Dry Ginger (6.9,0.2)
	Cashew Kernel (29.8,1.6)	Urad (33.1,0.4)	Poultry Chicken (5.6,1.7)	Others (6.8,0.2)
	Tea (27.3,0.5)	Ragi (32.4,0.1)	Powder Milk (4.5,0.8)	Pickles (5.7,0.1)
	Apples (27.2,0.4)	Masur (25,0.2)	Biscuits (4.4,1.4)	Condensed Milk (5.4,0.2)
	Unblended Tea Leaf (22.3,0.7)	Mustard (24.2,0.2)	Maida (3.6,1.9)	Papad (4.6,0.3)
	Cotton Seed (21.3,0.5)	Groundnut (23.9,0.2)	Food Powder (3.1,0.6)	Mixed Spices (4.4,0.3)
	Bajra (20.8,0.5)	Sapota (21.8,0.2)	Cabbage (3,0.8)	Bread and Buns (3.6,0.3)
	Mutton (20.5,1.4)	Unblended Tea Dust (21.1,0.4)	Okra (2.1,0.5)	Fish Meal (3.6,0.2)
	Dry Chillies (19.3,0.7)	Cardamoms (18.9,0.1)	Cauliflower (1.7,1)	Grapes (3.2,0.4)
	Cashwenuts (19.3,0.7)	Roasted Cashew Nut (17.7,0.1)	Rice Bran (1.6,0.5)	Pork (2.8,0.2)

(*Table A1 continued*)

(*Table A1 continued*)

	High inflation		*Low inflation*	
	High weight	*Low weight*	*High weight*	*Low weight*
	Ghee (17.3,0.9)	Pineapple (15.9,0.2)	Gram (–0.5,1.4)	Salt (2.8,0.2)
	Blended Tea Leaf (15.2,1.1)	Sweet Potatoes (15.6,0.1)	Betelnuts, Arecanut (–2.1,0.4)	Wheat Bran (2.7,0.4)
	Cattle Feed Gola (14.7,0.5)	Canned Fish (15.6,0.2)	Palm Oil (–2.9,1.7)	Ice Cream (1.5,0.2)
	Milk (14.3,13.3)	Vegetables Seeds (14.8,0.4)	Mustard and Rapeseed Oil (–3.3,1.9)	Chire Beaten Rice Flakes (0,0.2)
	Rice (13.7,7.4)	Ginger (14.7,0.2)	Tomatoes (–4.3,1.1)	Barley (–1.7,0.1)
	Blended Tea Dust (12.1,0.6)	Cummin (14.4,0.2)	Vanaspati (–5.1,2.9)	Black Pepper (–2.8,0.1)
	Maize (11,0.9)	Garlic (13.7,0.3)	Mangoes (–6.4,2.7)	Copra Oil (–3.1,0.4)
	Others (10.8,0.8)	Coffee (11.5,0.3)	Groundnut Oil (–6.7,1.3)	Besan Gram Powder (–4.5,0.4)
	Papaya (10.7,0.4)	Butter (11.4,0.3)	Coconut (–7.2,1)	Gingelly Oil (–4.6,0.2)
	Wheat (10.6,4.6)	Canned Meat (10.7,0.1)	Soyabean Oil (–7.5,1.6)	Lemon (–14.6,0.3)
	Green Peas (10.5,0.5)	Bagasse (10.4,0.2)	Brinjal (–9,1.2)	Corriander (–22.8,0.1)
	Marine Fish (8.8,3)	Cakes and Muffins (9.9,0.1)	Sunflower Oil (–12.2,0.7)	Guava (–28.7,0.3)
		Sugar Confectionary (9.6,0.2)	Rice Bran Oil (–14.4,0.8)	
		Rice Bran Extraction (9.3,0.4)	Cotton Seed Oil (–17,1.1)	
2010	Beef and Buffalo Meat (41.6,0.5)	Turmeric (128.6,0.3)	Powder Milk (8.5,0.8)	Bread and Buns (8.6,0.3)
	Marine Fish (40.5,3)	Garlic (100.2,0.3)	Bajra (8.3,0.5)	Besan Gram Powder (8.4,0.4)
	Inland Fish (37.8,2.4)	Cardamoms (89.4,0.1)	Blended Tea Leaf (7.8,1.1)	Sugar Confectionary (8.4,0.2)
	Mangoes (34.9,2.7)	Guava (76.6,0.3)	Wheat (7.8,4.6)	Unblended Tea Dust (8.3,0.4)
	Milk (24.6,13.3)	Litchi (74.5,0.2)	Rice (7.6,7.4)	Coffee Powder (7.9,0.1)
	Cauliflower (24.1,1)	Bagasse (47.9,0.2)	Maize (7,0.9)	Canned Meat (7.7,0.1)
	Poultry Chicken (23.7,1.7)	Moong (46.3,0.3)	Blended Tea Dust (6.9,0.6)	Lemon (6.3,0.3)
	Oranges (22.5,0.5)	Urad (35.7,0.4)	Cashwenuts (6.2,0.7)	Cakes and Muffins (5,0.1)
	Papaya (21.8,0.4)	Tapioca (29.4,0.3)	Cashew Kernel (5.8,1.6)	Barley (4.8,0.1)

Cabbage (21.7,0.8)	Black Pepper (22.4,0.1)	Unblended Tea Leaf (5.5,0.7)	Gingelly Oil (4.6,0.2)
Egg (21.7,0.8)	Rice Bran Extraction (20.9,0.4)	Vanaspati (5.3,2.9)	Condensed Milk (3.9,0.2)
Brinjal (20,1.2)	Butter (20.8,0.3)	Food Powder (5,0.6)	Salt (3.6,0.2)
Banana (19.2,1.4)	Sweet Potatoes (20.6,0.1)	Coconut (4.5,1)	Mustard (3.4,0.2)
Ghee (17,0.9)	Pineapple (20.5,0.2)	Green Peas (3.9,0.5)	Coffee (3.3,0.3)
Processed Prawn (16.4,0.5)	Pork (19.3,0.2)	Apples (3.7,0.4)	Roasted Cashew Nut (3.3,0.1)
Sugar (14.7,7.1)	Wheat Bran (18.2,0.4)	Dry Chillies (3.5,0.7)	Mixed Spices (2.2,0.3)
Groundnut Oil (13.5,1.3)	Sapota (14.6,0.2)	Sunflower Oil (3.2,0.7)	Ragi (2.2,0.1)
Okra (13.3,0.5)	Khandsari (14,0.3)	Soyabean Oil (1.8,1.6)	Pickles (1.6,0.1)
Tomatoes (13,1.1)	Groundnut (13.7,0.2)	Molases (1.6,0.5)	Copra Oil (0.3,0.4)
Onions (11.9,0.7)	Gur (13.3,0.3)	Cotton Seed (–0.6,0.5)	Maize Atta (0,0.1)
Mutton (11.9,1.4)	Cummin (13.2,0.2)	Palm Oil (–1.3,1.7)	Chire Beaten Rice Flakes (0,0.2)
Cattle Feed Gola (11.3,0.5)	Dry Ginger (12.3,0.2)	Mustard and Rapeseed Oil (–1.7,1.9)	Excluding Oil (–0.5,0.2)
Arhar (11.2,0.6)	Vegetables Seeds (11.9,0.4)	Rice Bran Oil (–1.9,0.8)	Canned Fish (–0.5,0.2)
Atta Wheat Flour (10.4,1.6)	Jowar (11.9,0.4)	Cotton Seed Oil (–2.6,1.1)	Others (–1.9,0.2)
Betelnuts, Arecanut (9.6,0.4)	Grapes (11.6,0.4)	Gram (–3.1,1.4)	Fish Meal (–5.7,0.2)
Biscuits (9.3,1.4)	Rawa Sooji (11.3,0.3)	Rice Bran (–4.8,0.5)	Masur (–6.4,0.2)
Maida (9.2,1.9)	Ice Cream (11.2,0.2)	Others (–7.9,0.8)	Corriander (–29.8,0.1)
	Ginger (9.7,0.2)	Tea (–17.7,0.5)	
	Papad (9.3,0.3)	Potatoes (–32.2,0.8)	

TABLE A2. Trade Policies in Edible Oils: 1994–2005

Date	Policy
April 1994	RBD palmolein placed on OGL with 65% import duty.
March 1995	All edible oils (except coconut oil, palm kernel oil, RBD palm stearin) placed on OGL with 30% duty.
July 1998	Import duty further reduced to 15%.
1999–2000	A 10% surcharge (S) added-Import duty-total 16.5%.
December1999	Duty on refined oils raised to 25% +10% (S). Also 4% special additional duty (SAD) on refined oils.
June 2000	Import duty on crude oils raised 25%(B)+10%(S), refined oils to 35% (B)+10% (S)+ 4% (SAD), Import duty on crude palm oil (CPO) for manufacture of vanaspati retained at 15%(B)+10%(S).
November 2000	Import duty on CPO for manufacture of vanaspati raised to 25% and on crude vegetable oils to 35%. Import duty on CPO for manufacture, other than of vanaspati, raised to 55%. Import duty on refined vegetable oils raised to 45% (B)+ 4% SAD. Import duty on refined palm oil and RBD palmolein raised to 65% (B)+ 4% SAD.
March 2001, as amended on April 26, 2001	Duty on crude oils for manufacture of vanaspati/refined oils raised to 75% except on soya bean oil, rapeseed oil, and CPO, at 45%, 75%, and 75% respectively. Import duty on refined oils, including RBD palmolein, raised to 85% (B) except for soya bean and mustard oil where it was placed at 45% (B) and 75% (B) respectively due to the WTO binding. Also 4% SAD on refined oils.
October 2001	Import duty on CPO and its fractions of edible grade, in loose or bulk form, reduced from 75% to 65%.
November 2001	Import duty on crude sunflower oil or safflower oil reduced to 50% up to an aggregate of 150,000 tons tariff rate quota (TRQ) subject to conditions. Duty on refined rape, colza, or mustard oil reduced to 45% up to TRQ of 150,000 tons subject to some conditions.
March 2002	Status quo on import duty structure maintained. Import of vanaspati from Nepal brought under SAD of 4%.
August 2002	SAD made nonapplicable on vanaspati imported from Nepal under TRQ.
April 2003	Import duty on refined palm oil and RBD palmolein reduced from 85% to 70% and SAD dropped.
July 2004	Import duty on refined palm oil and RBD palmolein raised from 70% to 75%.
February 2005	Import duty on crude palm oil and RBD palmolein raised from 65% to 80%, and that on refined palm oil and RBD palmolein from 75% to 90%.

Note: Lahiri Committee Report (2005), (B) – Basic Duty, (S) – Surcharge, (SAD) – Special Additional Duty.

Comments and Discussion

Kanhaiya Singh
National Council of Applied Economic Research

The subject of the paper is timely and very important. There are distinctly two parts of the paper. In the first part the authors try to link food inflation with nonfood inflation and make comments on how the monetary policy should be conducted. In particular, the linkages are used to make comments on the choices of the goal variable for the central bank, namely, core inflation or the headline inflation.

In the second part of the paper, authors carry out mechanical calculation of contribution of each of the food-related component of the basket of WPI and group them in four compartments, namely, high inflation, high weight; high inflation, low weight; low inflation, high weight; and low inflation, low weight. Having done that, five top contributors to food inflation, namely, milk, sugar, wheat and rice, edible oil, and fruits and vegetables are picked up for discussion. The discussion on the identified articles are structured to cover (*a*) description of the price series, (*b*) conjectures on short-term and long-term factors affecting the price of selected products with the support of extant literature on food inflation, and (*c*) government interventions. However, there is no quantitative analysis to support the conjectures about trade policy, fiscal policy, and monetary policy. The paper limits itself to simple description of the policy stances of the government, and restrains to take positions about what should be done in long run or short run with respect to the policy variables, in particular trade policy and fiscal policy. There is no attempt in the paper to explain the process of food inflation through a behavioral model.

The conclusion part of the paper has several issues, namely, exchange rate, working capital, monetary transmission, price puzzle, which are not properly analyzed and discussed in the main body of the paper and appear as standalone section. This has rendered the paper wanting of rigor and focus. Important policy issues such as choice of monetary target and interest rate policy require much more statistically sound and empirically grounded analysis, which is grossly inadequate in the paper. However, the paper does bring in several issues as food for thought and further research.

Some of the key findings and concerns of the paper are as follows: (*a*) food inflation is consistent driver of nonfood inflation and hence overall inflation. Therefore, headline inflation should be target instead of core inflation; (*b*) top contributors in food inflation are milk, fish, sugar, rice, etc.; and (*c*) "multiplicity of instruments (e.g., different forms of support prices, domestic subsidies, futures trading) and as well as the number of government agencies makes the task of controlling inflation a daunting challenge" (p. 52). Yet, at another place, authors find it useful to state, "combining both the macro and the micro analysis, a case can be made that in the context of a developing country like India, both the central bank and other branches of the government can play a meaningful role in tackling the inflation problem" (p. 49). Thus, the paper does not come out with clear stand on basic problems, which could otherwise give some direction to agriculture trade policy.

Given the objective of the IPF, it is important that papers must come out with at least one substantial policy recommendation based on serious data analysis. It is here that there are concerns which need to be flagged.

A. A major conclusion of the paper is based on the inference that food inflation is consistent driver of nonfood inflation and hence overall inflation. The pass through has been obtained from a simple bivariate regression presented in Table 4 without any element of dynamics. The coefficient of contemporaneous variable does not contain much information beyond simple correlation and in no way one can infer about the direction of relationship. In addition, the relative price of food to nonfood is not robust in explaining the aggregate inflation (see regressions 4 and 5 in Table 5) when money is included. Again, these regressions are not dynamic and, therefore, no causality can be established to find out driver. The casualty test is not transparent because results are not presented. Unless there is pure unidirection causality from food inflation to nonfood inflation, it cannot be called as consistent driver of nonfood inflation. Since authors have mentioned about the possibility of bidirectional causality, the conclusions is questionable. The problem here is not with the statement but certainly with the methodology which forms the basis of such conclusion.

B. Food inflation is a supply-side problem and if we agree with authors that food inflation drives nonfood inflation, then there is all the more reason to target nonfood inflation because the latter already contains the effect of pass-through. However, if nonfood inflation drives food inflation, then targeting headline inflation may be desirable. If there is a bidirectional causality then the same argument can be used to say target headline inflation, albeit with caution. However, it should be borne in mind that the key argument for adopting core inflation as target variable for monetary policy is the limitation of monetary policy in responding to transitory shocks, particularly those from supply side. Therefore, choosing and setting a target for monetary policy is a complex process, requiring more rigorous analysis.

C. Authors have calculated nonfood component by removing food as well as fuel from the wholesale basket. This means the nonfood being presented in the paper is basically core inflation. The rationale to remove fuel price from the nonfood inflation is not clear. In fact, fuel prices may come out to be driver of food inflation as well as manufactured goods inflation. This aspect has been completely ignored in the paper.
D. Another policy, which authors appear to favor, is about reverse linking MSP to inflation rate, which is mainly based on Basu (2011). It argues that MSP should be low to reduce procurement in bad (high-inflation) years and should be high in good (low-inflation) years to procure more and stock up. This can be termed as insensitive policy, which treats farmers as manufacturers. It needs intensive analysis about the behavioral causes of inflation in food prices before thinking of such a policy. In case food inflation turned out to be a cost push and weather-affected phenomenon, the farmer would be double crushed.
E. Then the paper uses two other papers to strengthen the arguments based on income elasticity of demand for food items, which is possibly calculated by using consumption expenditure as proxy of income because there is no unit-level data on income. Such elasticity is always a problem because it is very difficult to call those equations a demand equation from where the elasticity is calculated, primarily due to the absence of price variables, while total expenditure is used as proxy for income. The very idea to proxy income by the expenditure itself is questionable.
F. Paper has dealt with case studies of selected food products with high contribution to inflation. Here long-run and short-run explanations of inflationary process are discussed, which can at best be called as hypothesis because there is no statistical analysis to back the long-run and short-run issues. In fact, the very classification of short-run and long-run explanations is ad hoc and arbitrary. The economic sense of long run and short run is better described by statistically significant relationships of variables in level or difference. It is also not clear whether authors are explaining long-term inflation or short-term inflation by using so-called long-term issues. It is important because most of the time discussions are concentrated around spikes in inflation series rather than the average inflation.

G. Authors have focused on micromanagement of inflation and relevance of monetary, trade, and fiscal policies because demand has been identified as the main reason of inflation. Supply-side issues appear to have been ignored. Therefore, it is in order to throw some light on grossroot problems and, to cut short, we select the case of milk only.

H. Milk appears to have the largest impact on food price inflation. It is argued that the income elasticity is driving demand, which is causing inflation. The NSSO reports since 1991 to 2009 indicate that the average share of milk and milk-related products in food consumption during 1991–2000 and 2001–09 has been 15.2 and 15.4, respectively, for rural sector and 18.6 for urban sector for both periods. Considering the fact that the relative price of milk is very high compared to other food items, the consumption of milk in physical units must have fallen given the rigidity in its share in the monetary terms. Therefore, argument that increasing real income is leading to increased real demand of milk (in physical units) prima facie appears to be misplaced. Also, social system, food habits, and role of milk in feeding children suggest consumption of milk to be inelastic. However, milk faces cost-push inflation, in particular contributions of inflation in animal feed such as oil cakes or fodder cannot be ignored. All the time milk dairies increase price, the pertinent reason given is rise in the cost of production of milk. The structure of milk industry also plays important role, which includes the structure of the animals that are producing milk, state-level policies and the distribution of milk production across states. However, all these factors are also linked to the planning process and the emphasis on dairy development. The plan expenditure on dairy development has not been encouraging during recent years. At 1993–94 prices, the annual expenditure on dairy development during 1992–97 plan was INR 177 crore, which came down to INR 22 crore for 1997–2002 plan period and INR 34 crore during 2002–07 plan period. During 2007–10 the expenditure was just about INR 24 crore. When animal husbandry is also included with the dairy development, the combined annual real expenditure is INR 243 crore, INR 89 crore, INR 204 crore, and INR 127 crore, respectively, for 1992–97, 1997–2002, 2002–07, and 2007–2010. In 2002 OF was envisaged, but it has not taken off and as a result, the real expenditure on dairy development has gone down drastically resulting in decreased production capacity. Inefficiency of microfinance and rural credit system with respect to dairy development is reflected in the poor rate of increase of buffalo population, which is just about 0.74 percent per year during 2003–07, and the overall increase in the milch animals during the same period was restricted to 1.34 percent. The delivery system of credit and overall economics of cattle-raising needs revisiting. It is important that policies actions are taken toward increasing potential output by creating additional capacity and efficient distribution network over medium-term horizon if inflation is to be controlled. From the point of view of analysis, therefore, it is important to focus on supply side story also in greater detail along with issues of trade and monetary policies.

T.N. Srinivasan
National University of Singapore

Thank you very much. You talk about inflation in many ways: first inflation in nominal prices of all commodities, then food inflation, milk inflation, oil inflation, this inflation, and that inflation, etc. This does not make sense. There is a conventional notion of inflation with which most of us would agree. It is the old-fashioned one in which usually inflation represents the rise in prices of all commodities at the same rate. That is what is understood by inflation, not inflation by particular components. For example, suppose WPI did not change at all in a year but the nominal food prices changed. WPI has not changed means that the nominal value of the basket of WPI did not change and as such there was no inflation in a conventional sense. But the fact that food prices rose suggests to me that the price of food has increased relative to the aggregate basket or if you measure it in terms of the nonfood, it is the relative price of food to nonfood that has increased. Relative price determination is a quintessentially microeconomic general equilibrium story. It has nothing to do directly with the nominal price of food and macroeconomics of inflation. I wanted to start out with laying this out.

My second point is about measurement. The current state of price indices in India is extremely unsatisfactory. Until the new unified CPI for India was available. A new CPI became available only in February 2010, there was no single CPI. The WPI has been used for a long time for measuring inflation, called as the headline inflation. You should know that WPI is neither a PPI nor a CPI. It is not even a whole sales price index and it does not include services at all. How the price data is collected for WPI is a horror story by itself. One cannot really use WPI for inflation measurement meaningfully. It is not as if no CPI series are available. In fact, there are four of them relevant to particular population subgroups.

There are problems with weights of commodities in WPI. You have rightly pointed that the weight of food in the WPI is much lower than its weight in the NSS data on household consumption expenditure. Normally since CPI is used everywhere else in the world as a measure of inflation, the weights in CPI would be more relevant for inflation measurement purposes, particularly of food inflation, than those in WPI. NSS household survey data is not generally used in the WPI. Angus Deaton has published a number of papers including one paper where he compared the implicit unit values from NSS with the price data in one CPI and found there was close correspondence. Your rejection of the available CPIs altogether and exclusive focus on WPI is questionable. You could have used the CPI and made the qualifications necessary for the fact that they refer to subsections of the population, not the entire population of India; I have no idea what population group WPI refers anyway. Whatever the price index, it measures the change in the nominal value of the weighted bundle of commodities in the index between the base period and the current period. You can convert the change in a purely arithmetical way into a growth rate, so you have a rate of growth of aggregate price index, growth of food price index, and so on. But using these growth rates as a proxy for inflation requires a definition of inflation. Now, as I said defining a uniform rise fall in all prices of all commodities as inflation (deflation) is ideal, but this strict definition is very rarely applied anywhere in the world. The ideal definition is relaxed in various ways, but these relaxations are problematic. I have already pointed out that growth in food price index relative to the aggregate price index is a reflection of change relative price of food and not necessarily of inflation.

What about the economics of inflation? Unfortunately, the microeconomics of relative price determination and the macroeconomics of nominal prices and inflation are not well integrated. Friedman once quipped that inflation always and everywhere is a monetary phenomenon. Whether you buy it or not, not including any features of the monetary sector is not sensible. After all the Reserve Bank every other day is hiking the policy rates presumably in response to inflation or so it claims. Other, such as economy-wide policies besides monetary policies are important for studying inflation. Second, you have to think through the fact that some of the policy responses you mentioned, e.g., trade policy, are contingent responses to the observed value of the price index so that today's value of the price index may induce a response that may reflect itself in tomorrow's price index. For example, a rise in policy rate may simply be an ex post response to an observed increase in current to value of the index over the previous period, interpreted as current inflation or alternatively to be ex ante response to anticipated to future inflation. These two have different implications for modeling and interpretation. Now, none of these policies, except those in the charts you put in for sugar, milk, etc., are reflected in the analysis. I do not know how you can say much about inflation without bringing them in.

Whether the changes in the of food index are driving the changes in the aggregate index is again a causal question. It is not a question of whether food price growth and aggregate price index growth just moved together, it is a question of which is the forcing variable, which is the responding variable if I interpret your question correctly, and there is no answer to that from a simple OLS regression of pass through.

You have done causality tests. Presumably you did not do Sims test also, but you did the Granger causality test. It is about pure temporal causation, it has nothing to do with the economics of it. I would have loved to see more of the economics of price movements in the analysis than just thee pure time series properties analysis. And, as for co-integration, it is true that the existence of a co-integrating relation is interpreted as the existence of a long run "equilibrium" relation between the variables. Even if that is an equilibrium relationship there is no indication from that relationship which variable is causing, which is forcing, which is responding among its variables. So, I do not know whether one could infer very much just from the cointegration properties. The trade policy in India has been notorious as you know when it comes to foodgrains in particular. We have had notorious examples of onion exports being banned at the time of election when their prices went up; cotton exports being banned to response to the domestic textile lobbing for cheap inputs of cotton when cotton prices went up; and so on. If the policy with respect to domestic price change is of such nature, then that also has to be considered in your analysis if you wish to understand the mechanism.

The last point is, as I said this morning, you would need on one side a well-specified model of relative price determination on the micro side, and on the other side, you need to have a well-founded model on the monetary side and these two have to be integrated meaningfully in a micro behavioral sense. I do not think there exists such a model. The so called macro models with monetary sector are all ad hoc, either they bring in cash in an advance constraint or some other such constraint to motivate the holding of monetary or nominal assets in a portfolio. There is no model which says how individuals should respond to nominal price changes, with a micro underpinning or foundation. In my view this way of going about analyzing inflation would be any more useful than a seat-of-the-pants judgment about what the Reserve Bank should do. Is the RBI changing in the policy rate, having full knowledge of what the rise in its policy rate would do to the economy, the lag between the change in interest rate and the effect on inflation? As far as I know, there is no empirical analysis by the RBI of the stability of the transmission mechanism and time lag between interest rate change and its ultimate effect on inflation.

General Discussion

Isher Judge Ahluwalia, who chaired the session, opened the discussion by noting that rapid growth during the previous decade had increased the demand for food grains and other food items such as milk and fruits. Diversion of grains toward bio fuels had added to the shortages by cutting supply. Finally, given large fuel imports of India, rising fuel prices had also contributed to inflation.

Kenneth Kletzer said that this was a very through and detailed paper. Focusing on the relationship between relative price movements and inflation, he noted that monetary policy truly mattered. If we thought about why is there any potential long-run relationship between changes in the relative price of food and overall inflation, it is because monetary policy is responding to that in some way or the other. If we are looking at that short-run relationship, we talk about both monetary policy and the price-setting process when there are nominal rigidities or sticky prices in the economy or more generally imperfect competition. In this respect, the paper did not quite get you to talking about policy in the way that it is done.

Ashok Lahiri made four points. First, he expressed a lot of sympathy for using headline inflation rather than core inflation. Core inflation is useful only when the other prices are transitory in nature. For example, fuel prices may rise to $114 per barrel, but may come down to $72 within a short period. So you want to take out the persistent part in which the core inflation captures. But what the common man experiences in his daily life is headline inflation. Second, Lahiri questioned the argument that rising incomes have brought the inflationary pressures. China also faces the same problems in terms of increasing demand for food, but Chinese inflation is barely above 5 percent. Third, Lahiri suggested that the author look at edible oil more carefully. There are two components of edible oil; one is refined oil, the other is crude. The custom duty on refined oil has been changed several times in response to a rise in edible oil prices. Finally, Lahiri reminded of the old monetarist versus structuralist debate. Structuralist had argued that there are structural factors that impinge on monetary policy. For instance, private credit demand might rise with rising fuel and food prices. Likewise, increased need for working capital for inventories, higher wages, and increased government's food subsidy may push up the demand for credit. So how food price inflation feeds into or creates pressure on monetary accommodation may hold the key to addressing the concerns expressed by T.N. Srinivasan.

Surjit Bhalla questioned that MGNREGA could have been a principal cause of inflation since it represented such a tiny transfer of purchasing power. He also took issue with Isher Ahluwalia that growth could have been the cause since growth had been under way for three decades while inflation had raised its head only in the past three years.

Vijay Joshi took exception to the authors' conclusion that domestic policies such as support prices needed to vary depending on the level of inflation. Typically, support prices have been raised even in inflationary times. Joshi opined that the entire food procurement and distribution system and its relation with the relative price of food required much deeper analysis in the paper.

Mythili Bhusnurmath followed up on the point made by Joshi noting that also linked with the MSPs was the buffer stocking pattern. The latter had a significant impact on the price of cereals because we have buffer stocks much above what we require and even when prices were going up, government kept adding to the buffer stock and was not willing to off-load from it.

In her response, Prachi Misra thanked all commentators including the discussants. She agreed with T.N. Srinivasan that we need a well-specified macro model that includes the monetary sector and has explicit micro foundations. But that has to be the agenda for future research. She noted, however, that the discussion had not provided guidance on what the best course was, given the available data. We do need to know what commodities are driving the overall price increase if we are going to understand whether inflation hurts the poor. She noted that the study was descriptive in nature and the policy implications were conjectural.

References

Anand, Rahul and Eswar Prasad. 2010. "Optimal Price Indices for Targeting Inflation under Incomplete Markets," IMF Working Paper no. 10/200.

Bai, J. and P. Perron. 1998. "Estimating and Testing Linear Models with Multiple Structural Changes." *Econometrica*, 66 (1): 47–78.

Ball, Laurence and N. Gregory Mankiw. 1995. "Relative-Price Changes as Aggregate Supply Shocks," *The Quarterly Journal of Economics*, 110 (1): 161–93.

Barth III, M.J. and V. Ramey. 2001. "The Cost Channel of Monetary Transmission". Working Paper, University of Californian, San Diego.

Basu, Kaushik. 2010. "The Economics of Food Grains Management in India." Available at www.finmin.nic.in/WorkingPaper/Foodgrain.pdf

———. 2011. "Understanding Inflation and Controlling It," *Economic and Political Weekly*, 46 (41; October, 8–14): 50–64 .

Birthal, Pratap Singh, P.K. Joshi, Devesh Roy, and Amit Thorat. 2007. "Diversification in Indian Agriculture Towards High-value Crops: The Role of Smallholders," IFPRI discussion papers 727, International Food Policy Research Institute (IFPRI).

Blanchard, Olivier. 2009. "The State of Macro," *Annual Review of Economics*, 1 (1): 209–28.

Bruno, Michael and William Easterly. 1995. "Inflation Crises and Long-run Growth," NBER Working Paper No. 5209, Cambridge, Massachusetts, National Bureau of Economic Research.

Catao, Luis and Roberto Chang. 2010. "World Food Prices and Monetary Policy," IMF Working Paper No. 10/161.

Cecchetti, Stephen. 2007. "Core Inflation is an Unreliable Guide," VoxEU, March 1, 2007.

Chand, Ramesh. 2010. "Understanding the Nature and Causes of Food Inflation," *Economic and Political Weekly*, 45 (9; February 27).

Chen, Shaohua and Martin Ravallion. 2010. "The Developing World Is Poorer Than We Thought, but No Less Successful in the Fight Against Poverty," *The Quarterly Journal of Economics*, 125 (4): 1577–1625 [MIT Press].

Christiano, L.J., M. Eichenbaum, and C.L. Evans. 1997. "Sticky Price and Limited Participation Models: A Comparison," *European Economic Review*, 41: 1201–49.

de Janvry, Alain and Elisabeth Sadoulet. 2008. "Methodological Note: Estimating the Effects of the Food Price Surge on the Welfare of the Poor." ARE, University of California at Berkeley.

Dohlman, Erik, Suresh Persaud, and Rip Landes. 2003. "India's Edible Oil Sector: Imports Fill Rising Demand," Outlook Report No. (OCS090301), November.

Easterly, William and Stanley Fischer. 2001. "Inflation and the Poor," *Journal of Money, Credit and Banking*, 33 (2; May): 160–78.

Edelweiss. 2010. "Sugar Report." India Equity Research. Available at http://xa.yimg.com/kq/groups/22151721/2045523989/name/Sugar+edel.pdf

FAO. 2003. "A Review of Milk Production in India with Particular Emphasis on Small-scale Producers." PPLPI, Working Paper no. 2. Rome.

Fischer, Stanley. 1982. "Relative Price Variability and Inflation in the United States and Germany," *European Economic Review*, 18: 171–96.

———. 1993. "The Role of Macroeconomic Factors in Growth," *Journal of Monetary Economics*, 32 (3; December 1993): 485–512.

Fischer, Stanley and Franco Modigliani. 1978. "Towards an Understanding of the Real Effects and Costs of Inflation," *Review of World Economics*, 114 (4): 810–33 [Weltwirtschaftliches Archiv, Springer].

Gaiha Raghav and Vani S. Kulkarni. 2005. "Foodgrains Surpluses, Yields and Prices in India," Global Forum on Agriculture: Policy Coherence for Development. November 30–December 1, Paris, France.

Gandhi, Vasant P. and Zhang-Yue Zhou. 2010. "Rising Demand for Livestock Products in India: Nature, Patterns and Implications," *Australian Agribusiness Review*, 18: 103–35.

Gokarn, Subir. 2011. "The Price of Protein," *Macroeconomics and Finance in Emerging Market Economies*, 4 (2; October).

Government of India. 2005–06. *Economic Survey*. Available at http://indiabudget.nic.in/es2005-06

Hume, David. 1752. "Essays: Moral, Political and Literary," published as: *Political Discourses*. Edinburgh: A. Kincaid and A. Donaldson. Available at http://www.econlib.org/library/LFBooks/Hume/hmMPL24.html

Jha, Brajesh. 2004. "India's Dairy Sector in the Emerging Trade Order," IEG Working Paper, Series No. E/243/2004.

Jha, Raghbendra. 2005. "Inflation Targeting in India: Issues and Prospects," ASARC Working Paper No. 2005/04. Available at SSRN: http://ssrn.com/abstract=870164

Jha, Shikha, P.V. Srinivasan, and A. Ganesh Kumar. 2010. "Achieving Food Security in a Cost-effective Way: Implications of Domestic Deregulation and Liberalized Trade in India", in A. Ganesh Kumar, Devesh Roy, and Ashok Gulati (eds), *Liberalizing Foodgrains Markets: Experiences, Impact and Lessons from South Asia*. New Delhi: Oxford University Press.

Kachru, R.P. 2006. "Agro-processing Industries in India: Growth, Status and Prospectus." Available: http://agricoop.nic.in.pdf

Kapoor, Mudit and Shamika Ravi. 2009. "The Effect of Interest Rate on Household Consumption: Evidence from a Natural Experiment in India," February 20. Available at SSRN: http://ssrn.com/abstract=1346813

Khan, Mohsin S. and Abdelhak S. Senhadji. 2000. "Threshold Effects in the Relationship between Inflation and Growth," IMF Working Paper 00/110, Washington, International Monetary Fund.

Kotwal, Ashok, Milind Murugkar, and Bharat Ramaswami. 2011. "PDS Forever?" *Economic and Political Weekly*, 26 (21; May 21).

Lahiri Committee Report. 2006. "Report of the Committee on Rationalization of Customs and Excise Duties on Edible Oils and Oilseed," Ministry of Finance, Government of India.

Mishra, Prachi and Peter Montiel. 2012. "How Effective Is Monetary Transmission in Developing Countries? A Survey of the Empirical Evidence," IMF Working Paper.

Mishra, Prachi, Peter Montiel, and Antonio Spilimbergo. 2012. "Monetary Transmission in Low-Income Countries: Effectiveness and Policy Implications," *IMF Economic Review*.

Mittal, Surabhi. 2006. "Structural Shift in Demand for Food: India's Prospects in 2020," Indian Council for Research on International Economic Relations (ICRIER), Working Paper no. 184, pp. 1–35.

National Statistical Commission (NSC). 2001. "A Report of the Committee on Price Statistics." Available at http://mospi.nic.in/mospi_new/upload/Final%20Report%20of%20theCommittee%20on%20Prices%20.pdf?status=1&menu_id=180

Panagariya, Arvind. 2005. "Agricultural Liberalisation and the Least Developed Countries: Six Fallacies," *The World Economy*, 28 (9; September): 1277–99.

Pattanaik, Sitikantha and G.V. Nadhanael. 2011. "Why Persistent High Inflation Impedes Growth? An Empirical Assessment of Threshold Level of Inflation for India," Reserve Bank of India, October.

Rajmal and Sangita Misra. 2009. "Transmission from International Food Prices to Domestic Food Prices: The Indian Evidence," RBI Staff Studies No. 6, August.

Rakotoarisoa, Manitra and Ashok Gulati. 2006. "Competitiveness and Trade Potential of India's Dairy Industry," *Food Policy*, 31 (3): 216–27.

Ravallion, Martin and Gaurav Datt. 1996. "How Important to India's Poor Is the Sectoral Composition of Economic Growth?" *World Bank Economic Review*, 10 (1): 1–25 (Oxford University Press).

Reis, Ricardo and Mark W. Watson. 2010. "Relative Goods' Prices, Pure Inflation, and the Phillips Correlation," *American Economic Journal: Macroeconomics*, 2 (3): 128–57.

Srinivasan, P.V. 2005. "Impact of Trade Liberalization on India's Oilseed and Edible oils Sector," Report prepared for IGIDR-ERS/USDA Project: Indian Agricultural Markets and Policy.

Srinivasan, T.N. 2008. "Price Indices and Inflation Rates," *Economic and Political Weekly*, 43 (26 and 27; June): 115–23.

Subbarao, Duvvuri. 2011. "Central Bank Governance Issues: Some RBI Perspectives," Comments, Governor, Reserve Bank of India at the meeting of the Central Bank Governance Group in Basel, May 9.

von Braun, Joachim, Ashok Gulati, Peter Hazell, Marie Ruel, and Mark Rosegrant. 2005. "Indian Agriculture and Rural Development Strategic Issues and Reform Options," IFPRI Strategy Paper.

Walsh, James P. 2011. "Reconsidering the Role of Food Prices in Inflation." IMF Working Papers, Number 11/71, pp. 1–21. April. Available at SSRN: http://ssrn.com/abstract=1802984

Zeileis, A. et al. 2003. "Testing and Dating of Structural Changes in Practice," Computational *Statistics Data Analysis*, 44, 109–123.

RENU KOHLI
Indian Council for Research on International Economic Relations

AGNES BELAISCH*
Threadneedle Investments

Do Capital Controls Matter in India?

ABSTRACT India's capital account liberalization began 20 years ago, but a lot of controls still remain. The central bank's strategy has been to open financial markets cautiously while ensuring risk management systems and competencies are in place to ensure the system's resilience to foreign shocks. Has the regulation insulated Indian markets from global shocks?

This paper estimates uncovered interest parity relationships and finds that capital controls continue to provide a certain degree of autonomy to the central bank when setting policy rates. Vector Autoregressive (VAR) models are constructed to derive impulse response functions illustrating the transmission of a shock to foreign capital inflows through the different segments of India's financial sector. To provide a sense of the complexity of the regulation on capital account transactions, the paper provides a primer on the controls in place, which it uses to identify coefficient restrictions in the Structural VAR model.

Keywords: *Financial Market Integration, Capital Account Liberalization, Financial Architecture, Spillover, India*

JEL Classification: *F36, G15, G18, O16*

Introduction

The world's investors have been eager to venture deeper into India's fast-growing economy. And Indian investors have been thirsty for more cross-border diversification opportunities. But entry and exit points in India's financial system have been limited. The stakes are high. Deeper financial integration would provide better opportunities for risk-sharing,

* Affiliations as of the July 2011 IPF Confernce; both Renu Kohli and Agnes Belaisch were at the International Monetary Fund when this paper was started. This paper is the sole responsibility of the authors and does not represent the views of either past or present employers.

improve the allocation of capital, promote productive investment, and ultimately, strengthen the resilience of India's economic growth. However, further capital account liberalization also opens wider channels for the transmission of global instability to Indian financial markets.

Following the balance of payment crisis of 1991, the Indian government followed a strategy of cautious liberalization, biased in favor of nondebt creating flows. In addition, the RBI has been intervening on the foreign exchange market to mitigate the impact of capital flows on the economy. In 2006, the government set up a committee to undertake a comprehensive review of capital account liberalization to seize up opportunities from further financial integration (Tarapore, 2006). This committee recommended relaxing some of the existing controls on debt inflows. Will the current financial crisis result in a reconsideration of the intention to further capital account liberalization?

India's stock market is currently the most open segment of the domestic financial system. Equity inflows have been converging toward their level in other large, open emerging markets. They tripled between the start of the 1990s and the end of the 2000s. In comparison, debt inflows rose just ½ a point of GDP in the period. At its peak before the September 2008 Lehman collapse, stock market capitalization, at about $1 trillion, was the tenth largest in the world and the third in emerging Asia. It represented some 95 percent of India's GDP. Foreign participation on the stock market accounted for about 20 percent of trading in 2004. In contrast, India's bond market is mostly closed to foreign investors, who hold just about 1 percent of the public debt stock. The domestic debt market is neither very deep nor liquid.

A key question is whether India's discriminative approach to openness has arrested the transmission of foreign shocks affecting the stock market to the rest of the financial system. Global events have had a visible impact on India's stock market returns. This is particularly true for events originated in the United States and the Euro Area, which affect global risk aversion. The current economic and financial crisis is a striking illustration.

In this context, the objective of this paper is to assess the responsiveness of India's domestic financial markets to shifts in foreign capital flows entering India through its open stock market. The exercise may also help inform the debate on the course of reform with capital account liberalization. This paper is not an event study around a crisis period and differs from papers looking at international transmission mechanisms. Many papers examine the extent of contagion by looking at the rise in the correlation of local and foreign returns following a crisis compared to precrisis levels. Forbes and Rigobon (2002) provide a review of the literature in this area and conclude

FIGURE 1. FDI and Portfolio Flows (Gross, Percent to GDP)

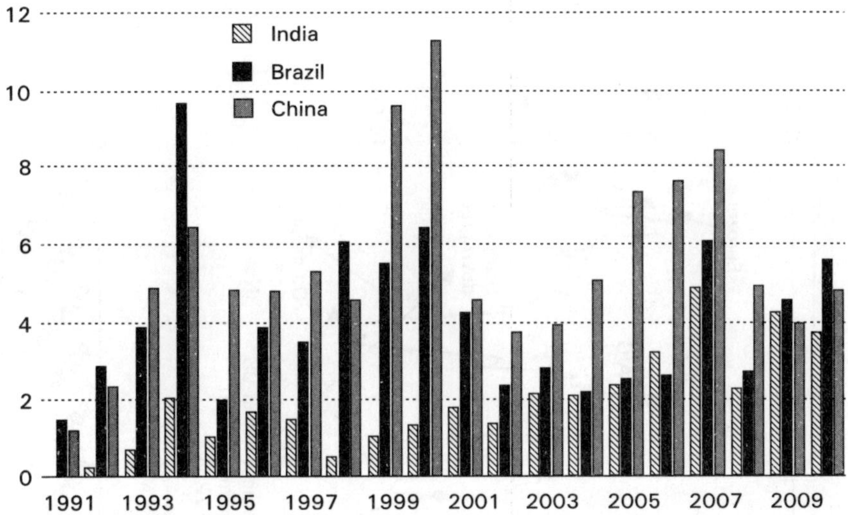

Source: Authors' calculations.

FIGURE 2. Capital Inflows to India (Gross, Percent to GDP)

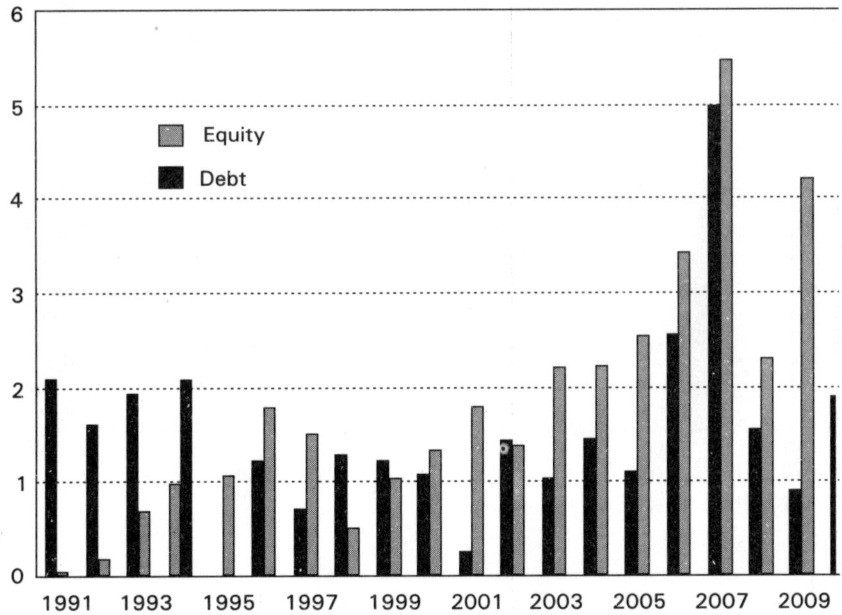

Source: Authors' calculations, drawing on the Zeileis/Shah/Patnaik dates of structural breaks.

FIGURE 3. Foreign Pressures on the Indian Stock Market (Sensex Index)

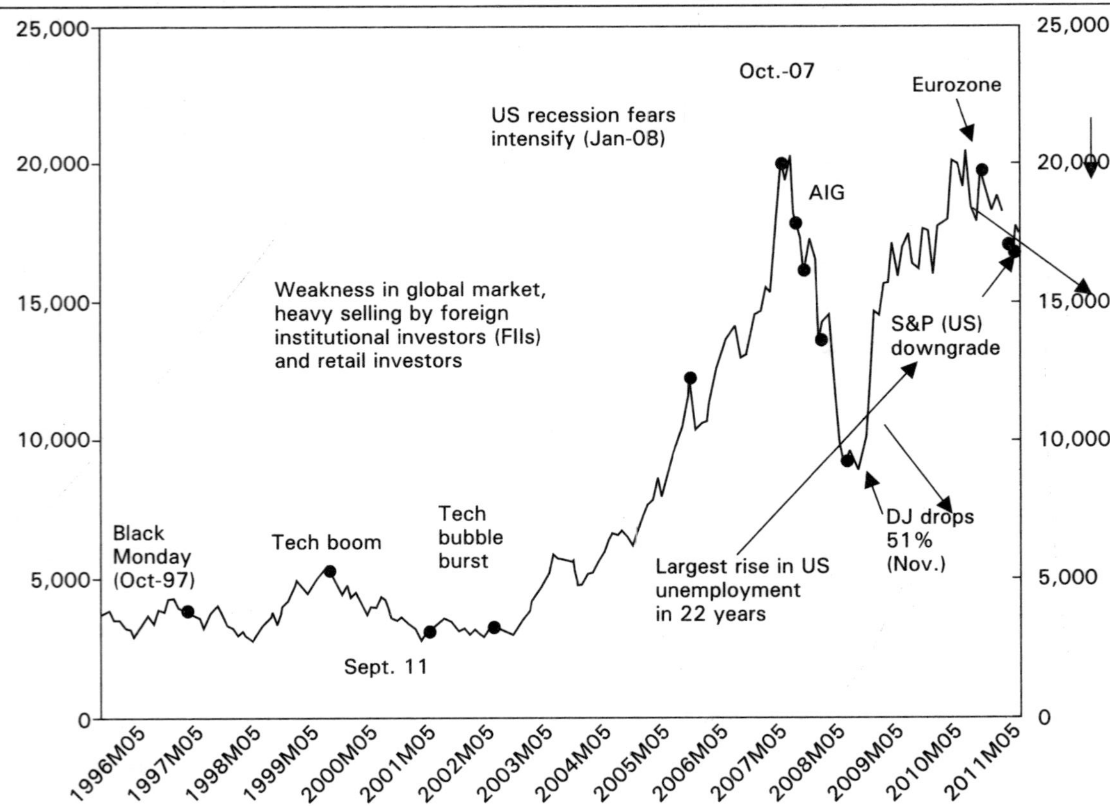

Source: International Financial Statistics and authors' calculations.

that if two markets show comovement during periods of stability, and if this continues—or increases—after a crisis, this does not necessarily constitute contagion but is rather evidence of interdependence. The approach followed in this paper is to identify the dynamics of propagation of a foreign shock to equity inflows to the domestic economy, looking in particular at whether it spills over to the closed segments of the financial system.[1]

We find evidence that capital restrictions on foreign participation in the Indian bond market insulate the bond market—but not fully. Arbitrage behaviors trigger a domestic channel of propagation between domestic financial markets. There is also evidence that central bank intervention temporarily dampens the exchange rate response to shocks to foreign equity flows. Partial sterilization leaves room for a response of interest rates, but with a lag. Inflation ends up higher than otherwise, but we find no significant impact on real variables. In sum, incomplete financial integration resulting from existing capital controls in the bond market provides a degree of short-term monetary autonomy to the central bank. Real economic activity appears relatively insulated from financial shocks. Regulation seems to have promoted the system's resilience.

The paper is organized as follows. Capital restrictions in India are complex and fastidious, and we take stock in section "A Primer on the Barriers to Capital Flows in India" of their composition at the start of 2011. Controls are disaggregated by type of flows (portfolio, debt, and foreign direct investment [FDI]), direction (inward and outward), and investors (individuals, banks, and corporations). The recent dynamics of liberalization is summarized to provide guidance on the government's intentions with this web of controls. To examine whether these controls are binding, section "Are the Controls Binding? Evidence from Uncovered Interest Parity" runs uncovered interest rate parity tests. It finds a persistent wedge between domestic and foreign interest rates adjusted for exchange rate differentials. Section IV examines whether regulation has been effective in terms of ensuring India's resilience to foreign shocks. We use VAR models to simulate the pass-through of foreign capital shocks to Indian markets, based on our understanding, from section "A Primer on the Barriers to Capital Flows in India," of how controls apply. And the last section draws conclusions.

1. This paper focuses on the financial channel of transmission and abstracts from spillovers through trade, which may also affect financial markets through exchange rate and output movements.

A Primer on the Barriers to Capital Flows in India

After successive reviews of the liberalization process in India, the capital account has remained not fully open and debt flows are tightly managed. Mohan (2008) gives a clear account of the logic behind this approach: the government considers that capital inflows in excess of the domestic absorptive capacity can lead to overheating in the economy and create asset price bubbles. Abrupt reversals of short-term debt inflows, in particular, can be detrimental to the real economy. Regulation discriminates against debt flows in two ways:

- Foreign investors' participation in the domestic government bond market is capped.
- Indian corporates can borrow abroad above a minimum maturity and below a maximum interest cost. These limits vary per economic sector, depending on perceived needs.

Restrictions on Inflows

By Issuers

The main route for foreigners to invest in India is to register as a Foreign Institutional Investor (FII) with the Securities and Exchange Board of India (SEBI). The status of FII was introduced in 1992, and there are currently some 1,700 FIIs registered.[2] Instead of direct registration, foreign corporates, individuals, institutions, or funds can open a subaccount with an FII, which can make investments on their behalf.[3]

- The FII status gives permission to buy and sell securities, open foreign exchange (FX) and rupee accounts, and remit and repatriate funds.
- Issuing a participatory note to an FII has been an alternate route of entry. The extent of foreign participation through this route varies with India's attractiveness; at the peak of the boom in June 2007, it

2. The following entities/funds are eligible for registration as FIIs: pension and mutual funds, insurance companies, banks, investment trusts, university funds, endowments, foundations, and charities can register; asset management companies, institutional portfolio managers, trustees, and power of attorney holders proposing to invest on behalf of broad based funds can also register as FIIs.

3. There were about 6,000 of them in November 2010.

accounted for 56 percent of all portfolio investments. In December 2010, it had fallen to 15 percent.[4]

- To take large stakes, a foreigner must request approval by the Ministry of Industry's Foreign Investment Promotion Board (FIPB) and the RBI, or only the RBI depending on the size of the investment and the sector in which the investment is to be made. Any foreign participation in financial services requires approval of the FIPB (e.g., to set up joint ventures (JV) in stock broking, asset management, or merchant banking).

Domestic investors are allowed to raise finance abroad, but with tight controls. Most transactions are permitted, but approval is needed to borrow above certain limits; there are price ceilings; and restrictions on end use of funds apply. Conditions are least restrictive for corporates and most restrictive for individuals, with banks an intermediate category. For the sovereign, external borrowing is mostly from multilateral and bilateral sources and on concessional terms. State governments are not allowed to borrow abroad directly.

Debt

The external borrowing of corporates is capped in amount, maturity, and cost. Beyond these caps, it is contingent on the specific approval of the central bank. There is an overall cap, set annually, which varies with macroeconomic conditions.

- Domestic corporates can borrow from abroad for commercial use up to US$500 million per year under the automatic route; those in the services industry can only do so up to US$100 million annually. Nongovernmental organizations engaged in microfinance can do so up to an annual US$5 million.
- Such external commercial borrowing (ECB) must have a minimum average maturity of three years for amounts up to US$20 million, and of five years for amounts between US$20–US$500 million.[5] Additional

4. This has become contentious in recent times, with the RBI recommending that participatory notes be banned (see Ministry of Finance, 2004). They were banned in 2007, in the wake of large, persistent capital inflows, but the ban was lifted in October 2008 in response to capital outflows during the global financial crisis. They are now restricted to 40 percent of assets under management.

5. These maturities have to be maintained in case of refinancing.

loans of up to US$250 million can be raised if they have an average maturity of 10 years.

- Funding cost must be below Libor plus 300 basis points (bps) for maturities between three to five years, and Libor plus 500 bps for maturities greater than five years. These restrictions were temporarily removed in 2009 as a crisis response to the rise in credit spreads in international markets.
- Finally, the end use of these loans is restricted: external financing for real estate,[6] stock market investment, repayment of existing rupee loans, domestic acquisitions, on-lending, and use as working capital are forbidden.

Banks and nonbank financial institutions must seek approval to borrow abroad. End-use restrictions on the funds apply as for corporates. There is an exception for banks and financial institutions engaged in infrastructure financing,[7] export financing, and debt restructuring packages, in which case the resources borrowed abroad can be used to finance acquisitions, equity investments, or on-lending of loans. All classes of borrowers are encouraged to hedge the associated foreign exchange exposure.

Banks can borrow abroad up to 50 percent of Tier I capital or US$10 million, whichever is higher. The short-term portion of the funds (with maturity of a year or less) is capped at 20 percent of unimpaired Tier I capital. Amounts borrowed abroad under this "automatic" route, which, thus, do not require prior approval, have to respect open position and maturity mismatch limits (GAAP limits). These limits exclude (*a*) all borrowings in the form of subordinated debt placed by foreign banks' headquarters with their branches in India as Tier II capital; (*b*) capital funds augmented by issue of innovative perpetual debt instruments (IPDI); and (*c*) other foreign borrowing requiring the RBI's specific approval.

Short-term external borrowing is subject to severe quantitative restrictions. A US$20 million ceiling applies to import-linked short-term loans, per transaction, and for permissible imports. Trade-related foreign loans with

6. An exemption is accorded to the "development of integrated township" as a permissible end-use, which was withdrawn in May 2007 and subsequently restored in January 2009 as a crisis response measure.

7. In April–May 2010, such institutions can borrow abroad up to 50 percent of their owned funds under the automatic route, and beyond with RBI approval. Eligible borrowers—in the sea port and airport, roads, bridges, and power sectors for the development of new projects—can also refinance domestic currency loans through a take-out finance arrangement with prior approval and subject to conditions.

one-to-three-year maturity and up to US$20 million are allowed to import capital goods. Overall funding costs cannot exceed 200 bps over six-month Libor for credit of one-to-three-year maturities.

Since October 2004, all ECBs availed with the general or specific permission of the RBI can be converted into equity if a set of conditions is met:

- The activity of the company is covered under the automatic route for FDI (see later), or if the foreign lender has obtained the Indian government's approval for taking equity participation in the company.
- The foreign equity stake after the conversion is within the sectoral cap for foreign participation, if any.
- The pricing of shares conforms with SEBI guidelines, as well as the Competition Commission of India's (CCI).
- The equity participation complies with other relevant statutes and regulation.

Since September 2008, Indian companies can tap foreign exchange resources against the equity holdings of their listed group companies with prior approval of the RBI. The foreign currency-exchangeable bonds of five-year minimum maturity can be issued in any freely convertible foreign currency, at a specified exchange price;[8] the rate of interest payable has to be within the all-in-cost ceiling under the ECB policy. Resources mobilized cannot be utilized for capital market or real estate investments and must be retained and/or deployed overseas.

Foreign investors of Indian origin (nonresident Indians, or NRI) can hold deposits in Indian banks, for a minimum period of 1 year. (The deposits themselves are also referred to as NRI.) The RBI modifies the rate of reserve requirements on such deposits to regulate the size of inflows. Rupee deposits receive a more favorable treatment than those in foreign currency.

- For one-to-three-year maturities, the interest rate on repatriable rupee deposits is capped at the Libor swap rate for US dollars of corresponding maturities plus 100 bps; and below the Libor swap rate for the respective currency and corresponding maturity for both fixed and floating rate foreign currency deposits. Banks can hedge this exposure.

8. At the time of issuance, the exchange price of the offered listed equity shares must be at least an average of the highest and lowest price of the share in the two weeks preceding the decision to raise capital abroad.

- Three-year deposit rates are applicable if the maturity extends beyond three years.
- Banks are free to levy penalty for premature withdrawal at their discretion. Banks may also, at their discretion, levy penalty to recover the swap cost of deposits. Where premature deposit withdrawal takes place before completion of the minimum stipulated period of a year, in which case no interest is payable, banks may at their discretion levy penalty to cover the swap cost. If the depositors are not informed of the penalty provisions at the time of acceptance of deposits, the exchange rate loss arising out of premature withdrawal has to be borne by the bank.

Bonded Debt

Foreign investment in domestic bonds is restricted. It is only allowed for foreign investors registered as FII. Until recently,[9] there was a ceiling on debt instruments at 30 percent of total investment, including units of domestic mutual funds, for FII participation in the Indian bond market; the remaining had to be held in equity. Setting up a 100 percent debt fund requires specific approval from SEBI. Their investment is capped at US$10 billion for government bonds and US$20 billion for corporate bonds; US$5 billion of these must be invested in bonds of minimum five-year maturity. In addition, 100 percent debt funds are restricted from buying unlisted corporate debt (this does not apply to the mixed fund category above). Since December 2006, FIIs can invest each year an incremental five percent of total net issuance in the previous financial year in government bonds. FIIs can also invest in debt capital instruments of domestic banks subject to an overall cap of 49 percent and individual cap of 10 percent of each issue for Tier I capital and the corporate debt limits in the case of Tier II capital.

Portfolio Investment

Portfolio investment in India was liberalized early, although some limitations remain.[10] First, it is restricted to approved FIIs. Also, an FII can only take equity participation up to 24 percent of its capital, as mentioned previously, subject to the following restrictions:

9. In October 2008, SEBI removed the 70:30 restriction on the share of investment in equity and debt, to give flexibility to FIIs in portfolio allocation.

10. Portfolio investment is defined in India in line with international standards, as an equity participation that represents less than 10 percent of a company's capital.

- A single FII can invest up to 10 percent in any company. Investment by firms and individuals registered as sub accounts of an FII is restricted to 5 percent. There are no restrictions on the repatriation of profits. Short-term gains (below a year) are taxed at 10 percent; longer-term gains are tax-exempt.
- Total holdings of all FIIs/subaccounts put together cannot exceed 24 percent in any company, but can be increased to the sectoral cap/statutory limit (as discussed later), as applicable, with the approval of its Board.
- Sectoral caps apply to FIIs as a category of investors. There is a separate cap on FDI and a composite cap on FII and FDI. Private sector banks, telecom, and media have composite caps.

Indian companies can raise capital abroad through Global Depository Receipts (GDRs) and American Depository Receipts (ADRs). Indian companies listing on registered exchanges are free to access the ADR/GDR markets without prior approval. There is no ceiling either. However ADR/GDR proceeds cannot be used to finance investment in real estate and stock market at home.

Foreign Direct Investment

Until January 2009, FDI guidelines consisted of different regimes for calculation of indirect foreign equity, i.e., proportionate (telecom/broadcasting sectors), regulation (insurance), and equity caps (banking, aviation, information technology, etc.) with sector-specific guidelines. In January 2009, FDI guidelines were revised to collect all types of foreign investments, i.e., FDI, investment by FIIs, NRIs, ADRs, GDRs, foreign currency convertible bonds, and convertible preference shares—for the computation of indirect foreign investment. The distinction is now based upon "ownership" and "control" by residents or nonresidents to decide upon overall foreign investment limits, direct as well as indirect.

Since 2000, foreign investors can acquire a 100 percent FDI stake in all industries, except for a list. For investment in those sectors on the list, case-by-case approval by the FIPB is required, except in six industries where FDI is totally prohibited. FDI also requires official approval when the foreign investor has an existing venture in the same field (Table 1).

Derivatives Investment

A registered FII may participate in all exchange-traded derivative contracts subject to limits prescribed by SEBI:

TABLE 1. India—Foreign Ownership Restrictions

	Sectors	*Ceilings (in percent of paid-up equity capital)*
No FDI allowed	Retail trading	
	Atomic energy	
	Lottery; Gambling	
	Housing and real estate business, except for bridges. Development of townships,	100
	construction of residential/commercial premises, roads or bridges.	
	Some agriculture business (plantations except tea)	
	Trading in Transferable Development Rights Manufacture of tobacco/tobacco substitutes	
Less than 100 percent FDI	Atomic Minerals	
	Private banks 1/	
	Telecoms 2/	
	Satellites	74
	Single brand product trading	
	Air transport services	51
	Investing companies in infrastructure/services	
	Asset reconstruction companies	
	Cable networks	
	Commodity exchanges	
	Credit information companies	49
	Infrastructure companies in securities markets,	
	Petroleum refining by public sector firms	
	Defense	
	Insurance 3/	
	Oil refining	26
	Publishing news and current affairs	
	Radio 1/	
	Public sector banks 4/	20
Up to 100 percent FDI allowed	Everything else, including SEZ, nonbank financial companies	0

Source: Government of India, 2006.

Notes: For the "less than 100 percent FDI" category, the ceiling applies to aggregate foreign participation across investment types. Details are as follows: 1/ For FII and FDI; 2/ For the sum of FDI, FII, NRI, FCCBs, ADRs/GDRs, etc.; 3/ Not to operate but as joint venture; 4/ subject to Banking Companies (Acquisition & Transfer of Undertakings) Acts 1970/80.

- Individual stocks. The FII position limit in derivative contracts on a particular underlying stock, i.e., stock option contracts and single stock futures, is 20 percent of the market-wide limit up to overall limit of ₹0.5 billion.
- Index options contracts. The position limit on a particular underlying index is ₹2.5 billion, or 15 percent of the total open interest of the market in index options, whichever is higher. This limit applies on open positions in all option contracts on a particular underlying index.
- Index futures. The position limit for contracts on a particular underlying index is ₹2.5 billion or 15 percent of the total open interest of the market in index futures, whichever is higher. This limit applies on open positions in all futures contracts on a particular underlying index. Hedge positions are permitted. In addition, FIIs can take exposure in equity index derivatives subject to the following limits:
  - Short positions in index derivatives (short futures, short calls, and long puts) must not exceed (in notional value) the FII's holding of stocks.
  - Long positions in index derivatives (long futures, long calls, and short puts) must not exceed (in notional value) the FII's holding of cash, government securities, T-Bills, and similar instruments.
- Interest rate derivative contracts. The maximum notional value of gross open position of a FII in exchange-traded interest rate derivative contracts is US$100 million. FIIs may take additional exposure in exchange-traded interest rate derivative contracts equal to the book value of their cash market exposure in government securities. The position limits for a sub-account in near month exchange-traded interest rate derivative contracts is ₹1 billion or 15 percent of the total open interest of the market in exchange-traded interest rate derivative contracts.
- FIIs are not allowed to participate in currency futures.

Nonresident Indians may also invest in exchange-traded derivative contracts in India. The instruments must be approved by SEBI and financed out of rupee funds held in India on a nonrepatriable basis.[11]

11. Repatriable funds are up to US$1 million per calendar year and held on account of legacy, bequest, inheritance, or sale of property.

Restrictions on Outflows

Indian corporates can participate in portfolio investment or FDI abroad subject to restrictions and ceilings:

- They can enter JV or start wholly owned subsidiaries (WOS) abroad up to 400 percent of their net worth without prior approval; investments exceeding this limit are allowed with RBI approval in the energy and natural resources sectors. This ceiling includes participation through capital contributions, loans and 100 percent guarantees on behalf of the JV/WOS. The ceiling is not applicable if the investment is made out of balances held in Exchange Earners' Foreign Currency account of the Indian party or out of funds raised through ADRs/GDRs. Domestic firms can also invest abroad through a Special Purpose Vehicle.
- Proprietorship concerns and unregistered partnership firms in the exporting sector can set up JVs/WOS abroad with the prior approval of the RBI and subject to certain eligibility conditions. The same applies to registered trusts and societies engaged in the manufacturing, education, and hospital sectors.
- Indian companies are prohibited from investing in foreign firms that are real estate or banking. Firms investing in the financial services industry overseas must be registered with domestic regulators in India for undertaking such activity, profitable for three preceding years, obtain approval from regulatory authorities in both countries, and meet the capital adequacy requirements in India.
- Listed Indian companies can invest up to 50 percent of their net worth in shares, and bonds issued by listed overseas companies, rated not below investment grade by accredited/registered credit rating agencies. Listed shares or bonds in companies that hold at least a 10-percent equity participation in a listed Indian company and in rated fixed income securities. This applies to mutual funds as well, with an overall cap of US$3 billion.
- Registered mutual funds can invest upto US$ 7 billion in ADRs/GDRs of Indian and foreign companies, equity of overseas firms, as well as initial and follow-on public offerings for listing at recognized stock exchanges abroad. They can also invest in foreign debt securities in countries with fully convertible currencies, short- term and long-term debt instruments, money market instruments, repos (not connected to own borrowings), sovereign bonds and short-term deposits with overseas banks, and all of investment grade rating. In addition,

mutual funds can also invest in derivatives traded on recognized stock exchanges overseas for hedging and portfolio balancing with underlying as securities, units/securities issued by overseas mutual funds registered with overseas regulators. Few can also invest up to a cumulative US$1 billion in overseas Exchange Traded Funds. And domestic venture capital (VC) funds registered with SEBI can invest in equity and equity linked instruments of offshore VC undertakings, subject to an overall limit of US$500 million.

- Indian companies may retain funds raised abroad through ADRs/GDRs for any period to meet their future foreign exchange requirements. Furthermore, resident shareholders of Indian companies can offer their shares for conversion to ADRs/GDRs and can receive the sale proceeds in foreign currency subject to the approval of the FIPB.
- Resident entities having overseas direct investments can hedge their exchange rate risks from such investments.
- Foreign corporates operating in India are free to repatriate benefits, after fulfilling tax requirements. The profits of foreign companies are taxed 41.8 percent, higher than the 33.7 percent rate that applies to local corporates. The tax rate on dividend and interest income for foreign companies, at 20 percent, is also higher than the 14 percent rate applying to local companies.

There is little liberalization of capital outflows for individuals:

- Convertibility of domestic assets (bank deposits) by residents is not permitted.
- Domestic residents can open foreign currency accounts at banks subject to certain restrictions (e.g., using foreign exchange earned and/or gifts received from close relatives and repatriated to India through normal banking channels by resident individuals). Foreign exchange earnings can also be brought in from the export of goods and services and royalties. These accounts are noninterest bearing only.
- Rupees cannot be bought abroad.
- Remittances overseas are allowed, but limited to US$200,000 per year after declaration of purpose.
- There is no limit on individuals' investment in overseas companies if these companies (*a*) are listed on a recognized stock exchange; (*b*) hold shares of at least 10 percent in an Indian company listed on a stock exchange in India.

- Individuals are free to invest in rated fixed-income securities abroad.
- Individuals can take interest-free loans up to US$250,000 with a minimum of one-year maturity from close relatives abroad.
- Indian students are considered NRIs and can borrow unlimited amounts outside India.

Banks operating in India are subject to restrictions on their foreign investments:

- They are permitted to invest abroad in money markets, mutual funds, and foreign stock markets up to the limits approved by their respective boards, subject to the Banking Regulation Act (1949) and open position limits.[12] Such investments must have a residual maturity less than a year and at least an AA(–1) rating.
- They can borrow from overseas branches and correspondents up to 50 percent of the unimpaired Tier I capital as the close of the previous quarter or US$10 million, whichever is higher.
- Banks can invest nonresidents nonrepatriable foreign currency deposits abroad in long-term fixed income instruments of maturity below that underlying these deposits, and subject to the securities having a rating of at least AA–.
- They can extend fund/nonfund based facilities to Indian JVs and WOS abroad up to 10 percent of unimpaired capital, and make rupee loans to NRIs with repayment debits to NRI accounts.
- They can open offshore banking units subject to the Foreign Exchange Management Act (FEMA) regulations.

A Historical Perspective

The current capital account restrictions reflect the order and sequencing of capital account liberalization that India followed.[13] The stock market was the first to open to foreign investors in 1993. Several factors contributed to this choice:

12. This Act stipulates that the assets of any bank in India shall not be less than 75 percent of the demand and time liabilities held at the close of business on the last Friday of each quarter.

13. For a detailed account of the process of capital account liberalization in India, see Reddy (2000), Kohli (2005), amongst others.

- Unlike most developing countries, India had a reasonably well-developed equity market when liberalization began in the 1990s, due to historical developments and a series of measures taken in the 1980s that established a sound institutional base for intermediating foreign capital with relative ease.
- The 1991 balance of payments crisis reinforced the urgency to reduce foreign currency debt and reorient the country's external financing pattern from expensive debt-financing toward cheaper equity-financing, along with risk-sharing.[14] These factors effectively shifted the weight of liberalization toward foreign portfolio equity capital, as even FDI at the time depended critically upon other economic reforms, including the sustainability and certainty of economic policies to inspire the confidence of long-term commitment of foreign investors.

The objective to avoid another macroeconomic crisis has been guiding India's strategy as regards its capital account opening strategy since the early 1990s. This strategy, first articulated in the Tarapore Committee I in 1997 and restated in the Tarapore Committee II in 2005, aims at striking a balance between fulfilling the country's foreign capital needs while ensuring balance of payments sustainability. The reports clearly link the dismantling of key capital account transactions with the achievement of macroeconomic objectives, like inflation reduction, fiscal consolidation, and monetary–fiscal separation. Since such critically interdependent reforms are slow to come about, in contrast to relatively easier-to-implement financial sector reforms, capital account liberalization has been a slow, evolving process.

As a result, the cautious opening to foreign investors of the domestic debt markets echoes the country's slow pace of fiscal consolidation and strong dependence on the domestic banking system for financing at a reasonable cost. In the early stage of liberalization, India's large fiscal deficit, traditionally financed by the vast public sector banking system through "financial repression," ruled out the opening of the domestic debt market to foreigners before reforms like interest rate deregulation and a reduction in fiscal dominance was achieved. But foreign participation in the debt market has been increased incrementally to support public debt issuance needs, and while a fiscal responsibility framework is implemented.

14. The High-level Committee on the Balance of Payments (1993) suggested this shift from debt to equity. Foreign liabilities were then mostly in the form of NRI bank deposits—India counted on these to finance the external gap. These were not eliminated but guarantees on interest rates and exchange rate risks were gradually phased out and passed over to depositors.

The sequencing of capital account liberalization must be understood in this historical context, reflecting the management of macroeconomic and external vulnerabilities (Reddy, 2000).

- In view of the external financing gap, restrictions on capital inflows have been removed before those on capital outflows—in times of surpluses, outflows have been liberalized fairly fast.
- Nondebt creating inflows have been preferred to debt-creating inflows to minimize liquidity and interest rate risks for firms.
- Long-term borrowings have been favored over short-term ones to enhance the productive capacity of the economy and lessen liquidity and roll-over risks from a sudden reversal.
- Fear of capital flight and the nonreadiness to make the currency convertible can explain why nonresidents' flows have been liberalized more than those of residents. Resident transactions are relatively easier to monitor and can be readily restricted, should the need arise, without affecting credibility in international markets. On the contrary, nonresident transactions can be discouraged by sudden policy reversals, damaging sovereign credibility and creating regulatory uncertainty. Residents thus remain the most restricted investor category to date. Within this class, the removal of restrictions for firms has preceded that for individuals.

The pace of capital account opening—slow and gradual—has often been commented upon by observers. This is not specific to capital account liberalization alone; much of India's reform process is characterized by gradualism (Ahluwalia, 2002). When compared with that in major OECD countries, however, the pace of liberalization in India has not been that slow. From the adoption of the OECD code on capital movements in the early 1960s to the complete dismantling of controls in the 1980s, Japan, the United States, the United Kingdom, and Germany took an average 20–25 years to fully liberalize their capital accounts (Griffith-Jones et al., 2000).

Historical experience does not obviously make the case that India's choices are different from others. However, it is hard to argue with the evidence that financial openness remains limited in India today, particularly when comparing with many other emerging market economies. The Chinn–Ito index of financial openness, which measures a country's degree of financial openness,[15] accords a score of –1.13—the lowest, or the least

15. The index measures a country's degree of financial and was initially introduced in Chinn and Ito (Chinn and Ito, 2006). It is based on the binary dummy variables that codify the tabulation of restrictions on cross-border financial transactions reported in the IMF's

capital openness—to India in 2007, compared to 0.99 for Brazil, 1.18 to Indonesia, –0.09 to Malaysia, 0.14 to Philippines, and –1.13 to Thailand. It is doubtful, however, as to how good a proxy it is for India for the index is unchanged since 1970, except for a minor blip in 2000, despite such a large change in India's external openness (Mohan and Kapoor, 2009).[16]

Are the Controls Binding? Evidence from Uncovered Interest Parity

There is an abundant literature arguing for a limited effectiveness of capital regulation in avoiding transmission. Magud and Reinhart (2006) provide a review of the arguments. It is, in any case, difficult to estimate the net gain provided by an open capital account from less volatility in capital flows but costlier domestic finance. As regards India, Ma et al. (2004) find significantly higher interest rates domestically than abroad, with the spread between onshore and NDF-implied offshore interest rates ranging between 400 and 1,000 basis points during 1999–2003. This spread narrowed in recent years, to a range of some 200–500 basis points. The fact that the onshore–offshore rupee yield differential failed to equalize over time suggests its declining trend owes more to inflation convergence than a significant liberalization, or circumvention of, capital controls.

There is, however, some evidence that capital controls have been partly circumvented in India, so that their effectiveness may not be as strong as desired by the authorities. A channel through which this has happened is the over-use of promissory notes (PNs). These are instruments issued offshore by investors wishing to take positions on the Indian equity and debt markets beyond the regulatory limits, or trying to avoid costly registration and capital gains tax. Singh (2007) reports that the stock of PNs amounted to $90 billion in 2007; he suggests that they had an impact on stock market fluctuations at times, e.g., through inflows for margin calls.[17]

Annual Report on Exchange Arrangements and Exchange Restrictions (AREAER) (Chinn and Ito, 2008). The index takes into account the following four categories of restrictions: (*a*) presence of multiple exchange rates; (*b*) restrictions on current account transactions; (*c*) restrictions on capital account transactions; and (*d*) requirement of the surrender of export proceeds (Chinn and Ito, 2008). The authors have constructed the index for 182 countries for the period 1970–2007.

16. Mohan and Kapoor (2009) point out that likewise, indices for Thailand, Korea, and Philippines are also seen to suffer from inertia for extended periods. They attribute it to the fact that such studies and the IMF's AREAER view capital account openness as a binary event: either open or closed, when in fact it should be seen as a process.

17. In the dialectics between regulators and markets, the RBI has taken measures that closed this loophole in the course of 2009, but there may be others.

FIGURE 4. Onshore–Offshore Spreads and Policy Rates (Percent)

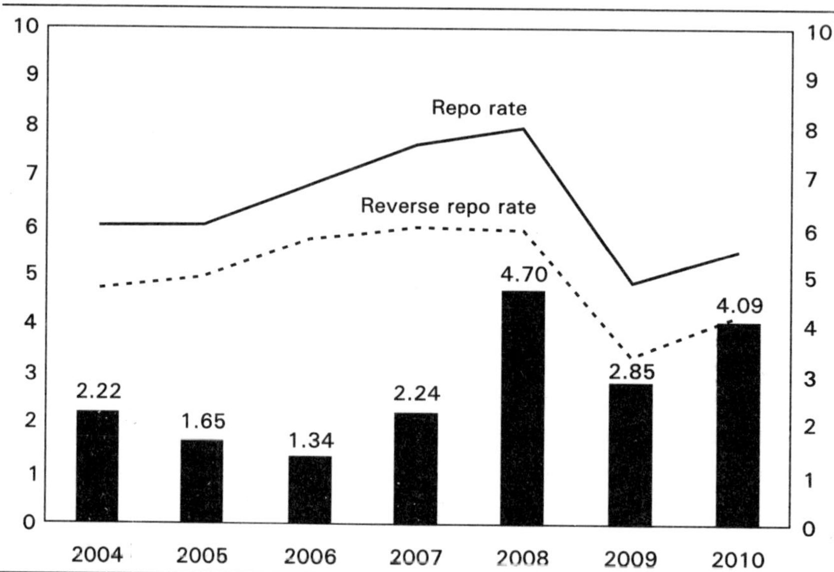

Source: Bloomberg and authors' calculations.

Note: 1/This is then spread between 1-year government paper and NDF (nondeliverable forwards) impled yields, calualted as Foward (NDF)/Spot* (1 + Libor)–1 for INR/US$ at1-year maturity.

The routes for domestic investors to arbitrage the interest rate differential at home and abroad, though growing, are still quite limited barring disguised current account transactions (or trade mis-invoicing). A popular route in recent times has been "overseas loans" by resident Indian firms. In the 2006–07 capital inflow surges, foreign capital entered India largely as portfolio flows (nonresident equity investments) and foreign loans taken by residents. The latter doubled to 3.4 percent of GDP between December 2006 and March 2007. They remained at these elevated levels until March 2008, when they came off sharply with the sudden deterioration in the external environment. Although some of these loans taken abroad financed the strong domestic investment growth, their sudden rise and reversal suggest they were possibly a leaky route for carry trade by residents. Domestic investors reportedly raised foreign currency loans abroad and deposited the converted rupees into domestic accounts through illegal routes.[18] The temporary rise in nondeliverable forward (NDF) trades during volatile market

18. There are other illicit routes for offshore–onshore arbitrage by residents. For instance, Indian promoters of publicly traded companies are reported to have raised personal loans offshore from large European banks against their equity holdings (these are worth at least $15 billion). The amount of such loans outstanding was reported to be $4–5 billion in 2007–08

conditions is also attributed to resident investors with trade-linked foreign currency exposures.[19]

The offshore market for NDFs has sprung up to circumvent controls, opening up new arbitrage opportunities. The active rupee derivatives market in Singapore is often cited as illustration, which allows investors to get an exposure to Indian interest rates and the rupee–dollar exchange rate, almost as they would onshore. Limited data are available on the volumes traded, that is, on the relative sizes of the offshore and onshore markets. This makes it difficult to gauge whether arbitrage is successful in eliminating the onshore–offshore interest gap—and whether, when convergence is obtains, it is a temporary or persistent phenomenon. BIS data highlights the growth of the offshore FX market (measured as the average daily turnover in dollar/rupee) between 2007 and 2010.[20] Gyntelberg and Remolona (2007) show that the rise in the offshore market turnover partly reflects leveraged trades, with short investment horizons and driven by carry-trade motives. However, trading activity is segmented due to market structure, indicating a limited potential of cross-border arbitrage. Nonresidents' participation is limited (it was 10 percent of daily onshore FX turnover in 2007). The share of financial investors in the total daily FX turnover, at 17 percent, is the lowest in Asia—the major driver of activity in the Indian currency market comes from exporters and importers (Tsuyuguchi and Wooldridge, 2008).

To test how biting capital controls are in India, we run tests of the UIP. The UIP condition postulates that capital mobility should equalize interest

---

and the money raised was brought in through an informal remittance system that operates outside traditional banking channels ("*hawala*") and through the accounts of nonresident Indians (see *Mint*, 2009).

19. The *Economic Times*, September 15, 2008 reported:

> The recent attack on the Korean won and its ripples across currency markets in Asia have turned out to be a money-making opportunity for players with overseas presence (...) Multinational banks operating in India, large corporates, and diamond houses have cashed in on the difference in the dollar-rupee exchange rate between the Mumbai currency market and the unregulated, unofficial, offshore markets in Singapore, Dubai and London (...) Corporates and institutions that have the means and flexibility have profited by buying the dollar in India and selling it on the offshore market, better known as the nondeliverable forward (NDF) market. Having undertaken such trades for years, many Indian corporates enjoy credit lines with banks abroad for NDF trades.

See Kohli (2009) for a fuller discussion.

20. In April 2010, it stood at $42.7 billion, twice the size of the onshore market (BIS 2010). In contrast, daily turnover on the offshore market was only estimated at $750 million in 2007, against $2–3 billion in onshore trading (2007 BIS survey).

rates across borders by arbitraging domestic and foreign returns expressed in the same currency.

The implication is that a regression of exchange rate returns on the interest differential should have an intercept of zero and a slope of unity. We acknowledge that this hypothesis has been consistently and decisively rejected in the data (see Engel, 1996, for a survey of the literature). Our goal is also to size the residual in the regression, often characterized as risk premium.

Estimations for either the full 2000–09 decade, or limited to its second half when controls were reduced, show that the UIP does not hold. The tests are run for short-term interest rate differentials, using both Indian and US overnight rates (the call rate and US federal funds rate) and three-month rates (Indian and US Treasury bill rates, as well as Mumbai interbank and US treasury bill rates). The rupee–dollar interest differentials are significantly different from zero for the full sample (Appendix table A.1). On average, the overnight and three-month interest differentials are in a statistically significant range between 480 and 313 bps over 2000–09 for two different measures of interest rates and for the two subsample sets. When we focus on the 2006–09 period, the result holds as well despite the rise in capital mobility.

We next examine whether deviations from the mean are transitory (before capital flows eliminate some of the arbitrage opportunity) or permanent (the result of historical barriers to capital flows). In this goal, we assess the dynamics of UIP over time, or its stationary properties. If the interest rate differential converges to some nonzero equilibrium level determined by India's sovereign risk considerations, this differential should be stationary. Unit root tests strongly reject the stationarity of uncovered interest differentials (except for short overnight rates), indicating that deviations from the mean are permanent (Appendix table A.2, upper panel). This implies that capital mobility, the equilibrating mechanism that would restore the parity condition, is insufficient to lead to convergence. The result is confirmed if restricting the test to the most recent period.

Finally, we assess the speed at which the empirical interest differential converges to the nonzero "equilibrium" level of the differential. The empirical interest rate differential is the level of the differential that reflects capital account restrictions and, in some way, represents a tax on domestic investors since domestic funding costs are higher than external ones. The "equilibrium" differential represents the country risk premium. The speed of convergence between the two differentials should depend on the degree of domestic capital market development and the degree of capital mobility—the higher the restrictions to capital flows, the slower the speed of adjustment to equilibrium. Under the assumption that the interest differential follows a univariate process (Cheung et al., 2003), the closer to one the coefficient in the autoregressive (AR) process, the more biting the capital controls.

The regressions continue to support the result of imperfect integration (Appendix table A.2, lower panel). The interest rate differential follows a strong AR pattern, with up to four statistically significant lags for the three-month Treasury bill rate—although it is the most liquid instrument in the Indian term money market. The AR(1) coefficient is close to unity for three out of four of the Indian rates used in the study. The adjusted R-squares are above 75 percent in all cases (data available on request).

The estimated UIP residual amounts to about 430 bps during the "tranquil" first half of the decade and 600 in the more volatile second half. The average credit risk premium for India (using credit default swaps) amounted to some 70 bps and 250 bps in each period respectively, indicating that the estimated deviation from UIP likely reflects a number of costs to investing in India beyond a pure reward for credit risk. These findings indicate that capital controls make the deviation persistent and predictable, as markets are not able to arbitrage it away, despite any anecdotal means found to circumvent them. The results hold over the full sample and in the most recent period as well. In the latter period, however, the persistence of the deviation has diminished as lagged coefficients are less significant. While the capital account has been more open, controls continue to provide some degree of short-term monetary autonomy to the central bank when it changes interest rates.

The Financial Market Transmission of Foreign Shocks: Evidence from Vector Autoregressions

The absence of full cross-border interest rate arbitrage in India raises the question of the magnitude of the impact of shocks to foreign capital flows on the domestic economy, through their effect on interest rates. The question is whether the existence of controls on capital flows insulated India from the domestic transmission of financial shocks during the recent global financial crisis, given both the large outflows that took place when the crisis struck and the large inflows experienced when global liquidity-supportive measures were adopted. Was capital account regulation useful in this context?

There are a number of reasons why capital inflows may affect interest rates in India:

- The link between the capital account and the exchange rate is significant in a country running a current account deficit, since financially driven capital flows can help offset trade-driven capital outflows if they are positive, or amplify the issue if not.
- The RBI has stated that its interventions do not aim at keeping the exchange rate from being market-determined over time but at smoothing its volatility. If it buys dollars to limit rupee appreciation following capital inflows, it will try partly to sterilize the impact on domestic interest rates of its intervention by entering reverse repo operations for liquidity absorption (selling bonds on the secondary market). This may put upward pressure on interest rates, depending on overall liquidity conditions. This may smooth the downside impact of inflows on interest rates.
- RBI intervention may also create room for an impact of capital flows along the yield curve. For example, if a large amount of short-term debt securities are sold to sterilize FX intervention, this may raise short-term rates and flatten the yield curve. If longer bonds or hikes in reserves requirements are used to remove liquidity instead, long rates may rise (requirements are met with long-term bonds). This steepens the yield curve. Liquidity considerations are known to have a large impact on India's yield curve.
- Finally, foreign inflows may indirectly reduce interest rates because rising equity prices lead to wealth effects, which may increase domestic bond demand (for diversification or savings purposes), and thus bond prices.

The economic literature has much examined the relationship between international capital flows and growth (see, e.g., Kaminsky, 2005). There is a variety of channels through which capital inflows may affect the real economy, including by providing liquidity to capital markets and accelerating domestic capital accumulation. Investment and consumption spending are promoted by wealth effects. The rise in savings supports demand in bond markets. Policy responses, in particular in the management of these inflows, may impact inflation. Based on these premises, we simulate the propagation of foreign financial shocks through domestic markets by estimating reduced-form VAR models. While these models have their shortcomings, they have been extensively used to investigate transmission mechanisms across countries or markets. They take into account the dynamics of interactions and feedback between the variables, including the endogenous policy response to shocks via monetary policy and FX intervention. We include the industrial production index to track the real effect of foreign shocks.

Key to VAR models is the identification of shocks based on restrictions derived from economic theory and practical considerations. We use two model specifications:

- First, a standard VAR specification defines each variable as a function of its own lagged values, as well as those of the other variables, thereby capturing the endogeneity over time of foreign and domestic financial variables. We apply the Choleski decomposition—a recursive decomposition of errors—that assumes a specific ordering of the endogenous variables to obtain impulse responses to observe a stylized representation of the Indian financial market. We order the less-controlled stock market first as the most exogenous segment of the financial system. We look at the response of financial and real variables to shocks to this market to assess whether it acts as a channel of transmission of external shocks.
- A structural VAR model is estimated next. The contemporaneous interactions between domestic and foreign variables are restricted based on the strictness of regulations on foreign capital flow by segment of the Indian financial market, as identified in section "A Primer on the Barriers to Capital Flows in India."

Data Specification

We use monthly data from January 2000 to October 2010. The foreign disturbance is captured by a shock to net portfolio investments inflows (denoted *FII*) to India, scaled by GDP. Short-term bonds are represented by the Treasury bill rate (*tbr*). We also check for robustness using the overnight interbank market rate, or call rate (*call*). Long-term yields are those of government bonds (*GSec*). The Sensex index captures stock market returns. The exchange rate is noted *INR/USD*. The monthly index of industrial production (*IIP*) provides an indicator of real activity. Finally, a set of macroeconomic variables, including the Libor interest rate and oil prices, are included to control for common shocks that affect all variables simultaneously. Variables are expressed in logarithms, except for interest rate and FII. Data sources are provided in the Appendix.

Over the sample period, variables are stationary in first differences, except for the call rate and inflation which are stationary in levels (Table A-3). First differences are taken month-on-month to preserve as much information as possible on the short-term pattern of interactions between endogenous variables. This transformation is sufficient to provide stationary time series. Likelihood ratio tests suggest the inclusion of one lag for each variable.

We isolate two subperiods for the analysis, 2000–05 and 2006–10, to test for changes in the extent of transmission in the latest crisis period of 2007–09. The period from 2000 to 2005 is a period of relative "tranquility." Then, a surge in capital inflows in 2006–07 was followed by a sudden reversal, triggered by the collapse of Lehmann Brothers in September 2008. This later period is characterized by a marked increased in the volatility of portfolio flows and other financial market variables (Figure 5).

FIGURE 5. The Surge in Volatility in the Second Half of the Decade

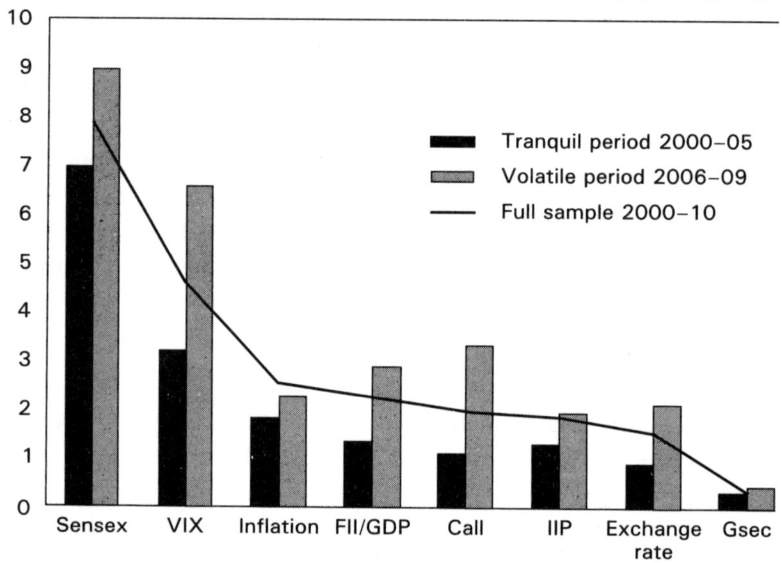

Source: Reserve Bank of India, Chicago Board of Exchange, and authors' calculations.

The models we use are symmetrical to a positive or a negative shock to capital inflows, and we simulate the impact of an unexpected rise in inflows. The positive one-standard deviation shock that we simulate represents a 2.3 percent increase in the ratio FII to GDP over the last decade. In the volatile period (2006–09), capital inflows display a standard deviation about doubles that in the more tranquil first half of the sample (2.9 versus 1.3). On average, foreign portfolio inflows represent 1.3 percent of GDP, and rise from 0.9 percent in the first half of the decade to 1.6 in the second half.

Stylized Facts from a Recursive VAR

We first apply the Choleski decomposition—a recursive decomposition of errors—that assumes a specific ordering of the endogenous variables to obtain impulse responses to observe a stylized representation of the Indian financial market. Such a model is unidentified: the errors are serially uncorrelated, but share common components across equations, so that a shock cannot be identified as originating from a particular variable. Identification is achieved by imposing and testing restrictions on the contemporaneous correlations of the variables at each time. The identifying restrictions imply a specific decomposition of these errors into orthogonal components, thereby recovering the original innovations driving the dynamics of the system.

Ordering

Financial markets are ordered by their degree of sensitivity to foreign capital flow shocks. As discussed in section "A Primer on the Barriers to Capital Flows in India", the Sensex stock market is the only significant recipient of foreign capital inflows, thus it is ranked first. The short-term money market comes next, as capital inflows impact liquidity conditions. The government bond market is ordered third. The exchange rate is ranked fourth. It is subject to central bank intervention, who explicitly aims at smoothing its volatility; its observed level may thus result from both foreign capital flows and the extent of intervention. Real activity is ranked last.[21]

Impulse Response Functions

The IRF show a much larger response to shocks to foreign portfolio flows in the volatile period than in the tranquil period (Figures 6.1–6.5). In the volatile second half of the decade, it appears that controls did not arrest the pass-through of foreign shocks:

- In the volatile period, the exchange rate appreciates at a faster pace and for a longer time than previously.
- The magnitude of the pass-through of the capital flow shock to the bond market is much smaller than to the stock market, the result of

21. By ranking real activity last, we handicap our simulations since we bias the regression against a contemporaneous impact of capital inflows on industrial production. Any result of the existence of an impact, thus, cannot be attributed to our ordering assumption but to pass-through in the system.

FIGURES 6.1–6.5. Impulse Response Functions to a Positive One Standard Deviation Shock to FII Flows

Impulse Response of Equity Returns to FII Shock

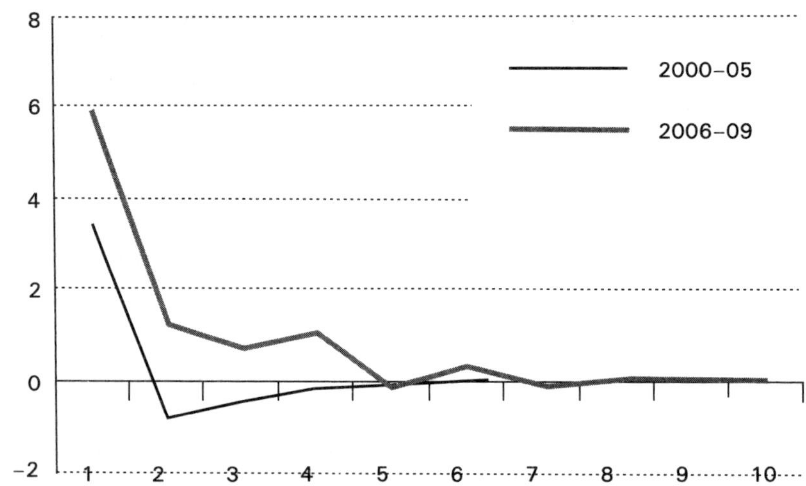

Impulse Response of Call Rate to FII Shock

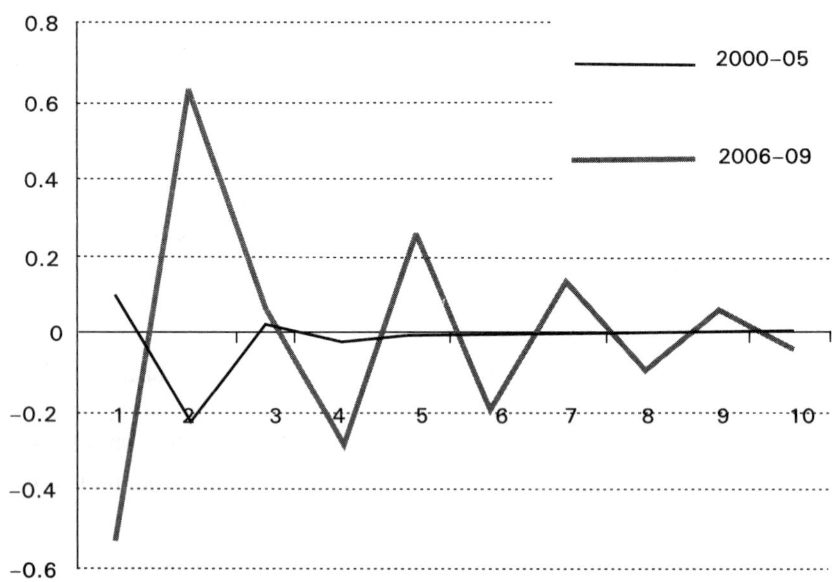

(Figures 6.1–6.5 continued)

(Figures 6.1–6.5 continued)

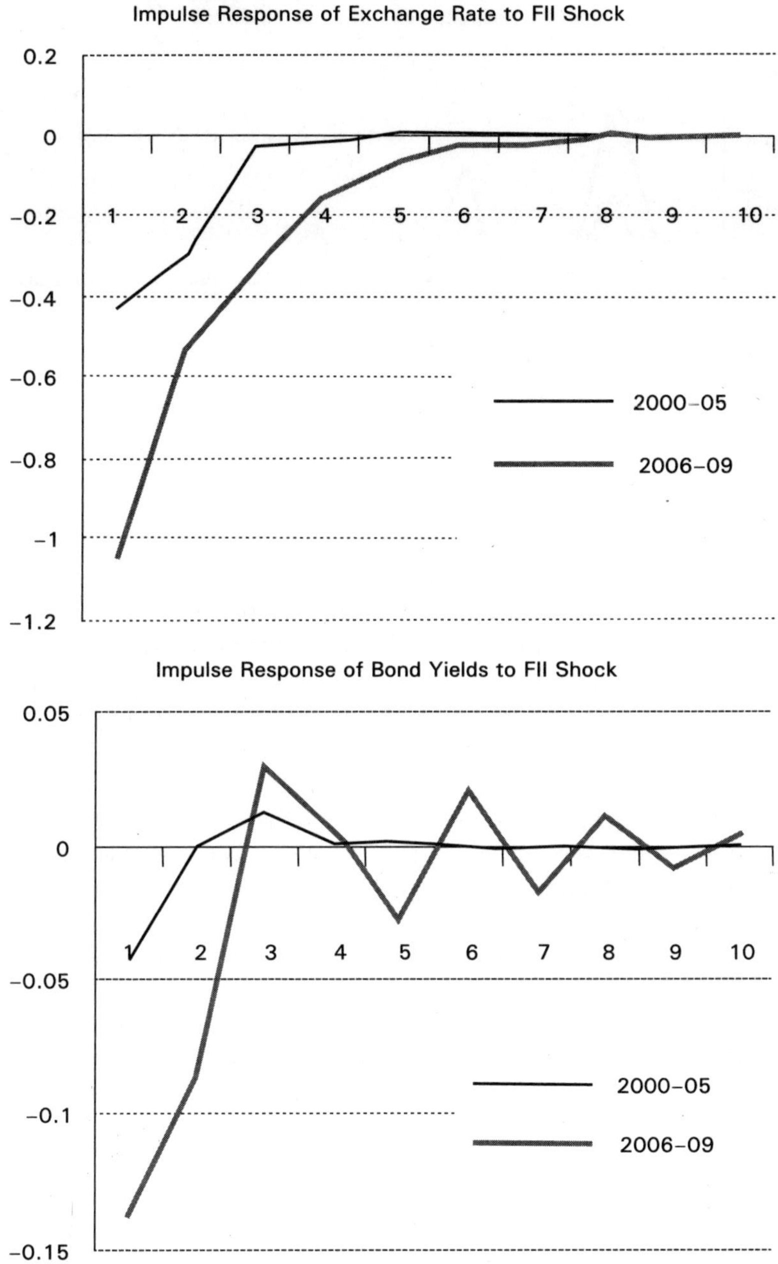

(Figures 6.1–6.5 continued)

(Figures 6.1–6.5 continued)

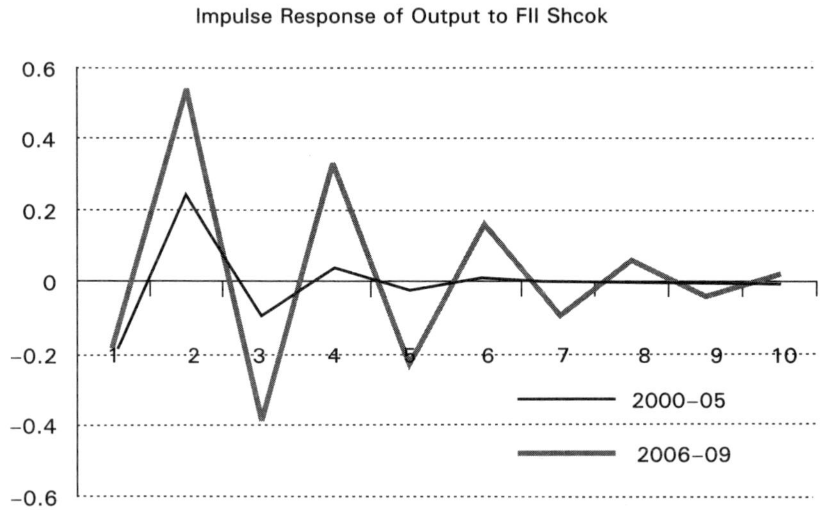

Source: Authors' calculations.

the presence of capital controls. However, the bond market spillover in the volatile period is large: it is more than proportional to the rise in the volatility of capital flows in the data.

- There seems to be a liquidity channel of transmission at play. Short-term interest rates and bond yields fall after the shock in the volatile period. The reasons cannot be identified by the model, but an RBI effect—a reduction of liquidity through the sterilization of FX operations, which should cause a rise in yields—is not visible. It is likely that the RBI did not fully sterilize its FX intervention at that time. In addition, stronger stock market performance may have resulted in a rise in bond demand by wealthier investors, for savings diversification purposes.
- The cumulative impact of the positive shock to capital inflows appears beneficial to activity. Activity is not insulated from the higher volatility; month-to-month volatility is large in the later period. It also takes longer to return to the steady state.

We check for robustness of the results by substituting the Chicago Board's index of implied volatility (or VIX) to FII inflows. This is a foreign shock truly exogenous to other VAR variables, as India does not, at least so far, influence global risk appetite. In this specification, the effect of innovations in the VIX on each of the dependent domestic variables may be better identified, given its low correlation with the other variables. This also reduces the order-dependent problem present in the Cholesky decomposition. To illustrate the equivalent of a positive shock to FII inflows, we simulate a negative one-standard deviation shock to the VIX. A fall in volatility is generally supportive of risk appetite, and equities tend to perform well. In all recent crises, from the collapse of LTCM to that of Lehman Brothers, this index spiked and risk assets sold off. The IRF confirm the results obtained in the previous specification and are presented in the Appendix for illustration (Figure A-1).

Structural VAR

A structural VAR (SVAR) model is estimated next, following Kim and Roubini (2000). We impose contemporaneous short-run restrictions on the relationship between endogenous variables, so that correlations can be interpreted causally with the help of identifying assumptions. The identification approach combines economic theory (e.g., Sims, 1986 and Bernanke, 1986) with a set of priors about the structure of Indian financial markets derived from the analysis in section "A Primer on the Barriers to Capital Flows in India."

The identifying restrictions are summarized in Table 2:

TABLE 2. Contemporaneous Interactions of Variables

Shock to *Effect on*	*FII*	*Sensex*	*Short-term rate*	*Exchange rate*	*Inflation*	*Long-term rate*
FII	1	X	X	X	0	0
Sensex	X	1	X	X	X	0
Short-term rate	0	0	1	0	0	0
Exchange rate	X	X	0	1	X	0
Real activity	0	1	0	0	1	0
Long-term rate	0	0	0	0	0	1

Source: Authors' tabulation with information from Ministry of Commerce, Government of India.
Note: X denotes a nonzero contemporaneous regression coefficient estimated by the multivariate model.

- FII flows are restricted to respond contemporaneously to shocks to stock market returns, the exchange rate and the short-term interest rate.
- Stock prices are assumed to react contemporaneously to all shocks, except long-term government bond yields.

- The exchange rate is treated as responding contemporaneously to innovations in foreign capital inflows, stock market performance, and inflation. However, given the absence of full arbitrage of domestic and foreign interest rate differentials evidenced in section "Are the Controls Binding? Evidence from Uncovered Interest Parity," the exchange rate does not respond to interest rate shocks.[22]
- We remove any contemporaneous response by the bond market to foreign innovations, given the restricted presence of foreign participation in sovereign and corporate debt market (section "A Primer on the Barriers to Capital Flows in India").
- Real activity is assumed not to respond contemporaneously to any shock in the model.
- The SVAR is overidentified. We can recover the structural shocks, in particular the capital inflows shock, using the IRF.

Over the full sample (2000–10), the likelihood ratio test for overidentifying restrictions shows that the null hypothesis that the restrictions are valid is accepted with an 8 percent probability of fitting the data. In the high volatility sample period (2006–09), the fit of the model improves significantly. The probability that the overidentifying restrictions are justified increases to 30 percent. These results suggest that the capital inflows shock in the SVAR is well-identified. The improvement in fit possibly reflects the deeper integration of global and domestic markets, leading to greater interaction between larger foreign capital flows and domestic asset prices. The still relatively low probability value is also a result of the lack of a sufficient period of estimation.

The IRFs provide the following results:

- In the full sample as well as the high volatility subperiod, both stock market and exchange rate responses to capital inflow innovations in the structural decomposition are higher than those traced in the Cholesky decomposition (Figures 6.6 and 6.7).

22. Investors do exploit small, short-term arbitrage opportunities. Foreigners usually have no incentive to trade the short-term rate onshore as the Offshore Interest rate Swap (OIS) curve provides the same return as the local MIBOR rate, but paid in US$. Differences between the offshore and onshore MIBOR floating rates would be arbitraged away by both local and foreign investors shifting liquidity to the most profitable leg, but onshore–offshore spreads can persist.

- The response of the bond market is more muted here. Bond yields fall, but even less in the volatile period than in the standard VAR specification. There may be domestic factors at play contributing a more significant role in determining bond yields, such as government deficit and the high food price-driven inflation in this latter period.

FIGURE 6.6. SVAR—Impulse Response Functions to a Positive One Standard Deviation Shock to FII Inflows, 2000–10

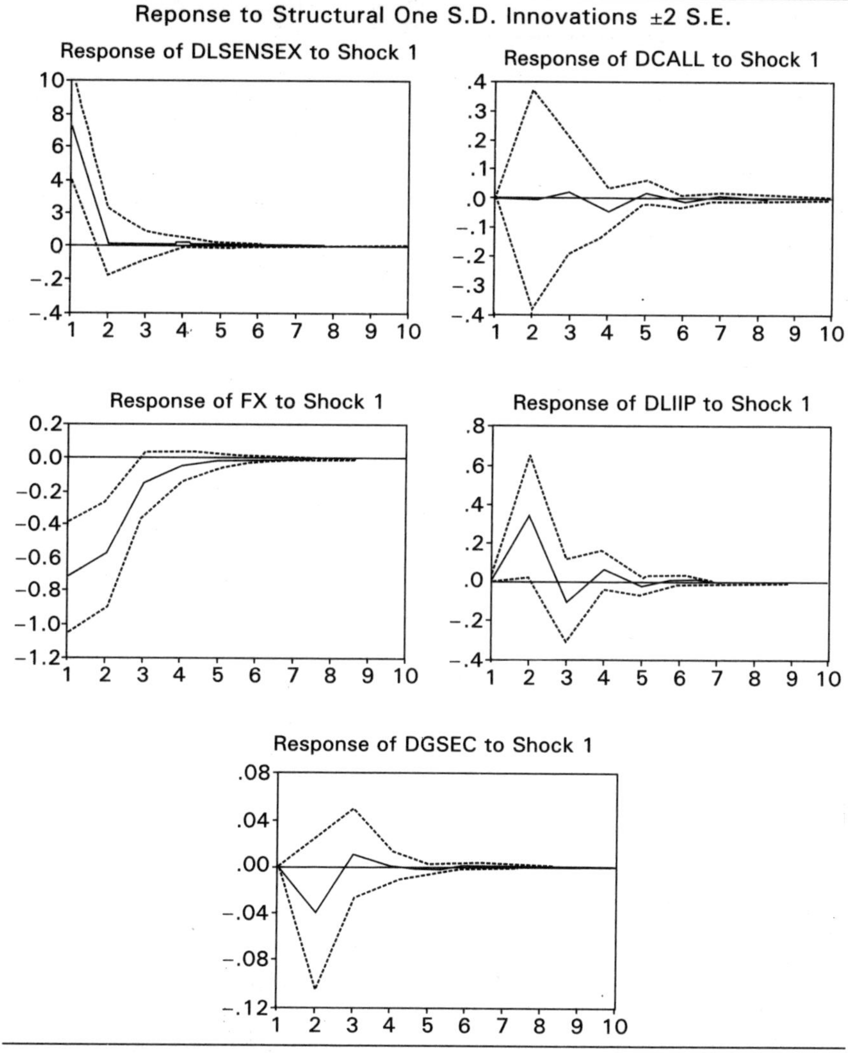

Source: Authors' calculations.

FIGURE 6.7. SVAR–Impulse Response Functions to a Positive One Standard Deviation Shock to FII Inflows, 2006–09

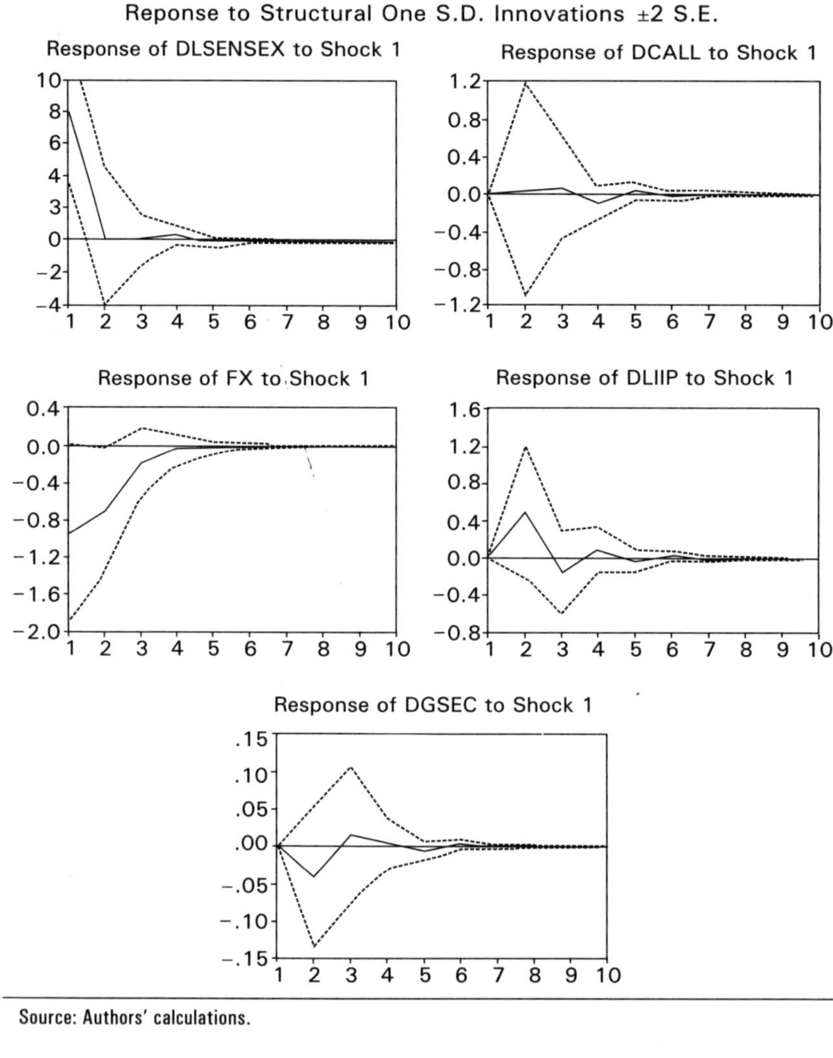

Source: Authors' calculations.

- Economic activity is impacted in the second month after a positive inflows' shock, rising and then falling to oscillate for two to three months before reverting to equilibrium. Output volatility appears to increase in the volatile period following a shock.
- Replacing FII flows with the VIX confirms the previous results. The impact of an external shock increases in the 2006–09 period of volatility.

Volatility seems to be the main result after the shock; it has increased in India over the course of the last decade. We decompose the source of this volatility according to the variables examined here, using the IRF to a shock to capital inflows in India. We find the following main results:

- Over the full period, shocks to equity capital flows have explained more than 25 percent of stock market volatility.
- Together with stock market volatility, shocks to equity capital flows explain about 60 percent of the dollar–rupee volatility.
- Shocks to capital flows are not associated with any interest rate volatility.
- In the second half of the decade, the impact of capital flow volatility on all variables has increased. In particular, capital inflows volatility now explains a quarter of the volatility in bond yields, despite the fact that bond market is closed to foreign investors.
- Over the whole decade, none of the variables examined here contributes much to volatility in real activity. This hides a change in the most recent period: interest rates mattered more for real activity (the policy rate moved a lot in both directions then), as well the exchange rate. This likely explains why it still matters today to the authorities to control FX volatility through some degree of capital flow restrictions.
- The impact of short-term rate changes on long yields has increased over the decade.

Conclusion

This paper points to the conclusion that capital controls have been effective in the period of relative tranquility that characterized the first half of the 2000s, but may have lost effectiveness in the second half, when volatility in capital flows was high. It is, of course, impossible to assess the counterfactual—maybe with no capital flow regulation, the pass-through to domestic markets would have been even higher.

There are other areas where the lack of counterfactuals makes a positive judgment difficult. The paper finds evidence that controls on capital flows do result in an unfavorable, positive wedge in the cost of finance at home and abroad. UIP estimations indicate that existing capital mobility fails to reduce the international interest rate differential. This incomplete financial integration resulting from remaining capital controls, especially in the bond market, nonetheless provides a degree of short-term monetary autonomy to

FIGURE 7. Variance Decomposition: Contributions to Volatility in 2000–10 and in the Turbulence Subsample 2007–09

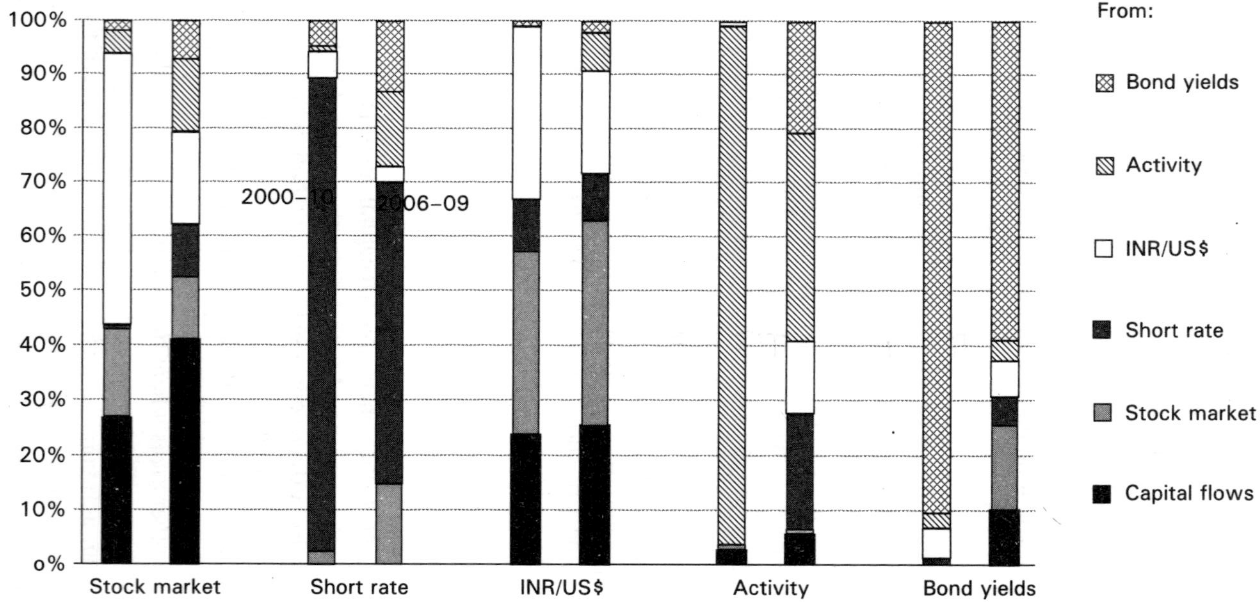

the central bank. These findings match those of Magud et al. (2011) who find that capital controls seem to make monetary policy more independent in their recent survey. The bulk of the empirical evidence on the effectiveness of capital controls also shows that the area where capital controls have been most successful is in providing more autonomy for monetary policy and altering the composition of capital inflows. Success on reducing the volume of inflows and exchange rate pressures has been mixed (Clements and Herman 2009).

As regards the channel of transmission of shocks to capital entering the stock market to the rest of the less open financial system, it appears that a standard liquidity channel is at play. Unexpected capital inflows appreciate the rupee exchange rate versus the dollar (the full impact is unobservable due to RBI intervention, and only the ex-post exchange rate is available). In the absence of full sterilization of FX intervention by the RBI, the net impact of capital inflows remains a rise in domestic liquidity. Interest rates fall. The impact is larger on government bond yields than on the short-term interbank rate. It is possible that investors save part of the increase in wealth from an unexpected rise in the stock market in government bonds. Capital flow and stock market volatility are found to explain a larger part of bond yield volatility in the recent period compared to the first half of the 2000s.

Real activity appears insulated from shocks to foreign capital flows shock. A similar conclusion is reached by the RBI in its recent analysis of the impact of the global financial crisis upon India (Report on Currency and Finance, July 2010). It is notable, however, that the direction of the response to a positive shock to inflows is itself positive. A stronger stock market and lower interest rates would certainly explain this effect. The exchange rate appreciation does not seem detrimental.

There may be some policy implications from these findings.

- Issuing domestic debt in local currency and limiting the entry of foreign investors on the domestic debt market during capital account liberalization, while keeping equity inflows relatively unrestricted, could be a useful pattern for other countries to study while liberalizing capital accounts. This combination appears to reduce volatility at home and allow monetary autonomy, while permitting the liberalized segment of the financial market to contribute to improving the pool of available capital at lower costs.
- Adopting a discriminative approach to the liberalization of different segments of the financial markets, and varying the pace of liberalization, can be successful in limiting the transmission of foreign shocks.

Financial liberalization could otherwise exacerbate vulnerabilities and threaten financial and macroeconomic stability. This, of course, has also a cost in terms of funding for investments.

- The increased intensity of transmission of global financial shocks in times of higher volatility in foreign capital flows may be partly attributable to India's rising financial integration, after some capital account regulation was relaxed in the mid-2000. However, the list of regulation remains long and complex, as evidenced by this paper, and the recent rise in the pass-through of foreign shocks points to the unavoidable need to strengthen financial markets infrastructure so as to improve resilience to shocks.

Appendix

FIGURE A-1. Robustness Check: Impulse Response Functions to a Positive One Standard Deviation Shock to Global Risk Appetite (the Inverse of the VIX Index of Volatility)

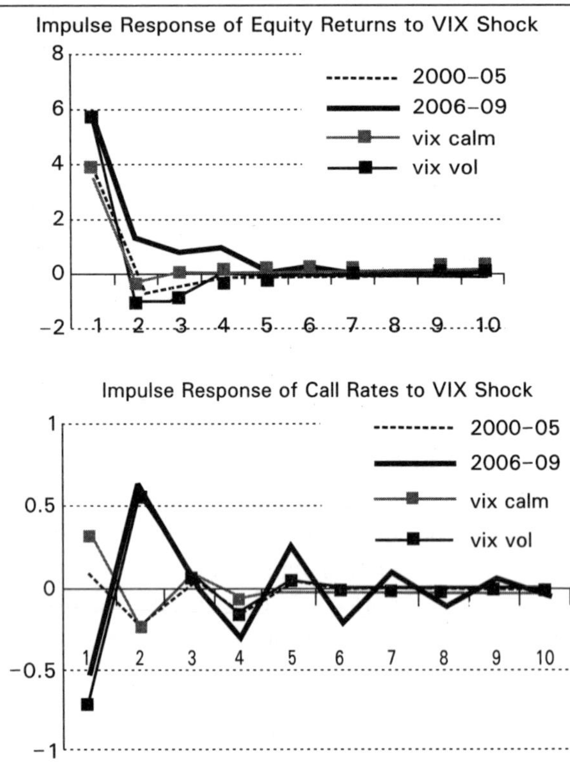

(*Figure A-1 continued*)

(Figure A-1 continued)

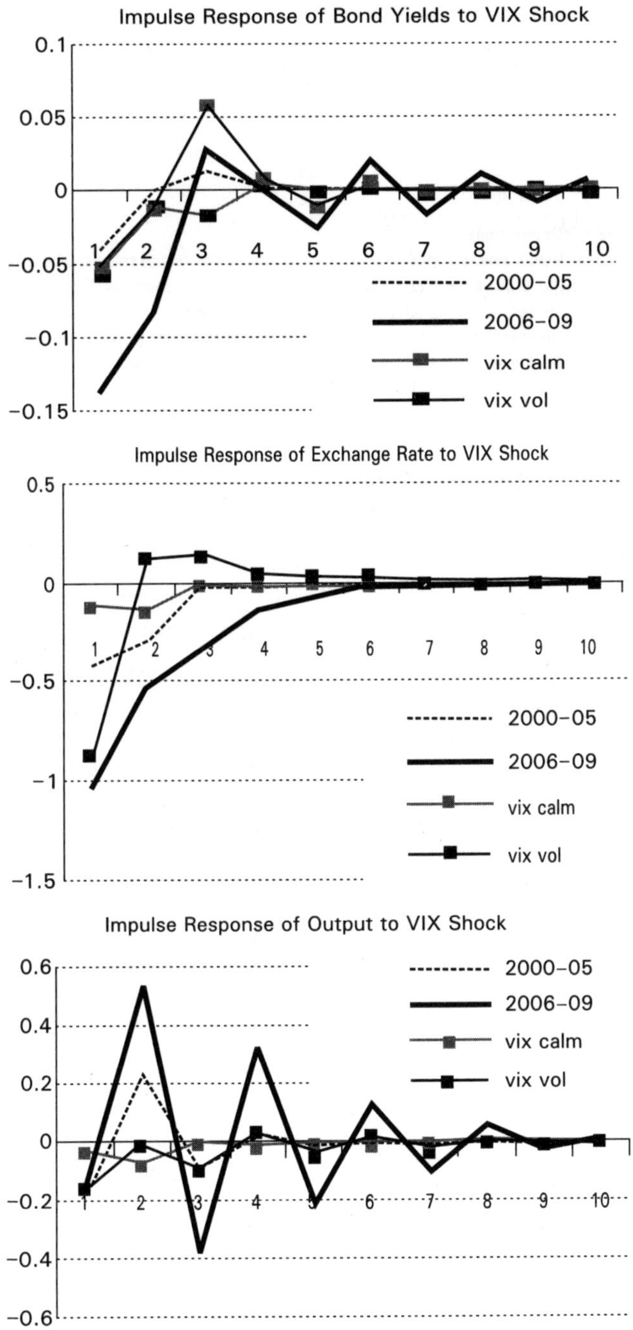

Source: Authors' calculations.

TABLE A-1. India–US Interest Rate Differentials, Significance Tests

	Average differential		*Trend*	
Overnight	*Full sample*	*2006–09*	*2000–05*	*2006–09*
Repo-FFR	4.73*	2.09**	–0.08	0.09**
Ca1l-FFR	3.25**	1.71	–0.13	0.04**
3-month Treasury bill	3.13**	1.95***	–0.03	0.09***
MIBOR	4.80*	2.51 **	–0.05	0.13***

Source: Authors' computations.

Note: Regressions report average uncovered interest differentials for full sample, 2001:01–2009:02, with dummy for 2006–09 to capture monetary policy shift in the period. Trend coefficients in columns 3 and 4 are intended to test for decline/rise in differentials in different regimes.

* , **, *** indicate significance at 10, 5, and I percent levels, respectively; regressions include AR(1) term, not reported Durbin-Watson test statistic lies in 1.96:2.26 range, indicating absence of serial correlation Adjusted R-square is 0.79–0.89 for all regressions.

FFR = US Federal Funds rate; MIBOR = Mumbai interbank overnight rate.

TABLE A-2. Time-series Properties of India's Uncovered Interest Parity $(i - i^* - \Delta S_t)$

	Repo-US FFR	*Call-US FFR*	*1-month Mibor-US Libor*	*3-month India-US TBR*	*3-month Indian MIBOR-US TBR*
	Unit root tests for stationarity				
2000–09	*No, I(1)*	*Yes, I(0)*	*No, I(1)*	*No, I(1)*	*No, I(1)*
2005–09	*No, I(1)*	*Yes, I(0)*	*No, I(1)*	*No, I(1)*	*No, I(1)*
Persistence 2000–09					
Ar (1)	1.36***	0.55***	0.79***	0.73**	1.05***
	(16.70)	(5.42)	(7.96)	(7.45)	(10.58)
Ar (2)	–0.23	0.03	0.11	0.18	–0.13
	(1.67)	(0.26)	(0.86)	(1.50)	(0.92)
Ar (3)	–0.38**	0.07	0.04	0.12	0.10
	(2.97)	(0.59)	(0.33)	(0.95)	(0.75)
Ar (4)	0.46***	–0.02	–0.20	–0.20*	–0.20
	(6.63)	(0.19)	(1.58)	(1.62)	(1.80)*
Ar (5)	–0.26***	0.10	0.16	0.09	0.17
	(6.52)	(0.91)	(1.57)	(0.85)	(1.59)
2005–09					
Ar (1)	1.16***	0.57***	083***	0.71***	1.06***
	(7.57)	(3.97)	(5.54)	(4.71)	(7.11)
Ar (2)	–0.77	0.10	0.13	0.26	–0.11
	(0.66)	(0.59)	(0.69)	(1.41)	(0.48)
Ar (3)	0.41	–0.02	0.13	0.19	0.11
	(1.56)	(0.16)	(0.66)	(0.98)	(0.48)
Ar (4)	–0.13	0.29	–0.28	–0.22	–0.26
	(0.48)	(1.54)	(1.32)	(1.14)	(1.14)

(*Table A-2 continued*)

(*Table A-2 continued*)

	Repo-US FFR	*Call-US FFR*	*1-month Mibor-US Libor*	*3-month India-US TBR*	*3-month Indian MIBOR-US TBR*
			Unit root tests for stationarity		
2000–09	*No, I(1)*	*Yes, I(0)*	*No, I(1)*	*No, I(1)*	*No, I(1)*
2005–09	*No, I(1)*	*Yes, I(0)*	*No, I(1)*	*No, I(1)*	*No, I(1)*
Ar (5)	−0.35* (1.86)	−0.33 (1.53)	0.05 (0.30)	−0.02 (0.88)	0.13 (0.74)

Source: Authors' computations.

Note: *, **, and *** indicate significance at the 10, 5, and 1 percent levels. T statistics are in parenthesis.

i is one of the following domestic interest rates. Repo denotes the Indian policy rate; Call is the Indian overnight money market rate; Mibor is the Mumbai interbank overnight rate; 3-month India is the Indian Treasury bill rate.

i star is one of the following foreign rates. US FFR is the Federal Funds rate; US TBR is the US Treasury bill rate. S is the INR/US$ exchange rate.

The Call rate equation includes a dummy for March 2007. Adjusted R-squares (not reported) range between 0.75–0.94.

TABLE A-3. Unit Root Tests (2000:01–2010:10)

	Level	*First difference*
VIX	2.61	9.64
FII_GDP	-9.62	
Sensex (log)	2.33	10.27
Call rate	9.19	13.98
GSec rate	2.26	10.04
INR/US$ (log)	2.29	7.68
IIP (log, s.a)	2.27	12.78
Inflation	6.46	
RBI intervention	7.78	
TBR	2.37	12.94

Source: Author's calculations with data from RBI.

Note: Augmented Dickey-Fuller tests reject the presence of a unit root in the variables in first differences, as well as in inflation in level, at the 5 percent significance level.

TABLE A-4. Data

Variable	*Source*
Call rate	CEIC Data Company
Mibor (3-month Mumbai Interbank Overnight Rate),	
Libor (USD Deposits, 1 month)	
Indian Treasury bill rate (3 month, Secondary market)	
Indian Government securities yields (1 year, 10 year)	
US Federal Funds rate	
Sensex Index (BSE Sensitive 30)	CEIC Data Company
Index of Industrial Production	CEIC Data Company
Exchange Rate (INR/US$)	CEIC Data Company
NDF rates	Bloomberg Database
VIX (Chicago Board of Trade)	Bloomberg Database

Comments and Discussion

Ajay Shah
National Institute of Public Finance Policy

What Is the Interesting Question? In thinking about this field, it is useful to start with an analogy. Recall the time when India had trade barriers. Suppose an economist asked the question: Does this drive a wedge between the Dubai and Bombay price? Suppose the researcher built up the datasets, and found that this holds for TV sets, but not for gold or diamonds.

This finding would have been interesting in saying that controls were not effective for a few commodities: gold and diamonds. For most commodities, the controls were effective. There is little doubt that trade barriers interfere with goods arbitrage.

The main body of international economics did not focus on the fact that hair dryers cost more in Bombay than in Dubai. The fact that hair dryers cost more in Bombay than in Dubai did not mean the controls were wise.

The really important questions that deserve focus are: Why did policy makers introduce the controls in the first place? What was their goal? In the trade context, this may have uncovered explanations such as an infant industry argument. This would have motivated an examination of the extent to which the controls catered to these goals, and the costs that were involved in pursuing such goals.

India's Missing De jure Capital Account Liberalization The Chinn-Ito measure, which is the most widely used database of de jure openness, reports that India's controls have been unchanged, with a score of –1.18, from 1970 onward. Most countries have moved forward with greater capital account liberalization than India, as measured by the Chinn-Ito score. In 1970, half of the countries of the world were as closed as India or worse. In 2009, one-third of the countries of the world were as closed as India or worse. India is increasingly an outlier, by world standards, on the extent of capital account restrictions.

With this very limited opening, what are we going to find in the data? We're likely to find big wedges between onshore and offshore prices. This has been observed in previous work by Sergio Schmukler, Robert McCauley, etc. If, in some respects, we find that onshore and offshore are integrated, we're going to be surprised.

How Integrated Are the Onshore and Offshore Markets? Figure 1 shows the onshore and offshore INR/US$ 90-day forward exchange rate. What is remarkable in this graph is the extent to which the two time-series move closely together. This suggests that despite substantial de jure capital controls, a substantial extent of arbitrage is taking place. This data is inconsistent with the assertion in the Kohli and Belaisch paper that the onshore–offshore rupee yield differential failed to equalize over time.

FIGURE 1. Onshore and Offshore INR/US$ Exchange Rate, 90 Days Forward

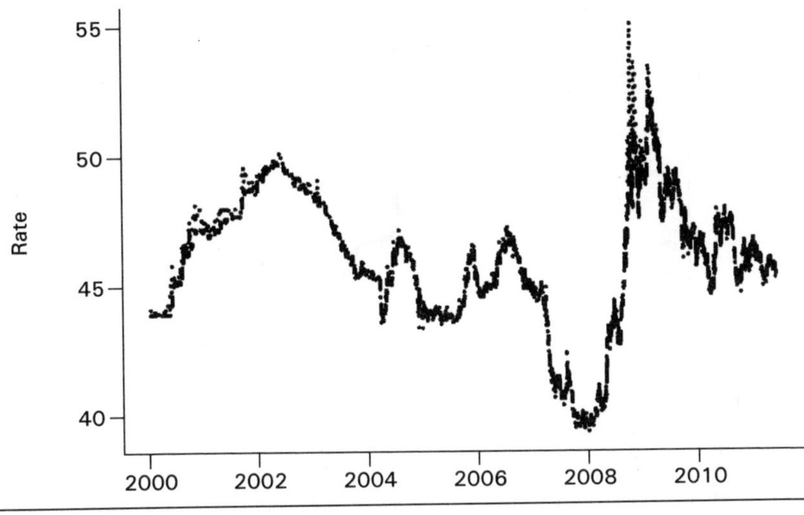

The Interesting Question Why did India employ capital controls? The major macroeconomic policy goal was to get to the corner of the impossible trinity where a pegged exchange rate was rendered compatible with freedom of maneuver for monetary policy. We should judge the capital controls regime by the extent to which it delivered on the goal of upholding the exchange rate regime (see Patnaik and Shah, 2011).

Figure 2 shows the evolution of the INR/US$ exchange rate volatility. We started out with a period of 4.74 years with a de facto pegged exchange rate, with an annualized volatility of 2.31 percent. The capital controls were

FIGURE 2. Evolution of INR/US$ Volatility

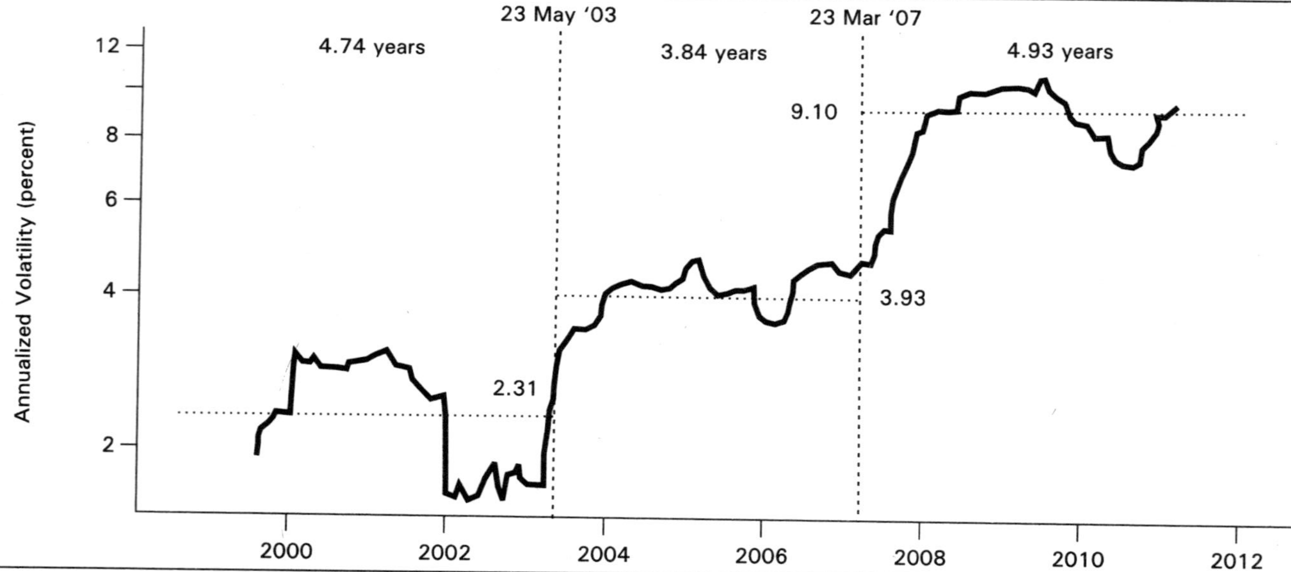

not effective enough for upholding this, and exchange rate volatility went up to 3.93 percent. This lasted for 3.84 years. Once again, the capital controls were not effective enough for upholding this, and exchange rate volatility went up to 9.1 percent, i.e., almost a floating exchange rate. This exchange rate policy framework has been the most durable one: it has lasted a full 4.93 years.

The key point is that India's system of onerous de jure capital controls were not able to uphold exchange rate pegging: India had to move to the floating exchange rate anyway.

On a side note, the presence of these structural breaks in the exchange rate regime implies that the estimation of the VAR of the Kohli and Belaisch paper needs to be done within each subperiod separately.

Summary India is one of the most closed economies of the world. It is not surprising or interesting that capital controls have driven a wedge between onshore and offshore prices. What is remarkable is the extent to which onshore/offshore arbitrage is effective, despite the prevailing capital controls. The really interesting question is: Did the capital controls deliver on the goals that they were intended for? With the abandonment of de facto exchange rate pegging in 2003 and then 2007, the answer seems clear: the capital controls were porous enough that pegging induced a substantial loss of monetary policy autonomy.

Mythili Bhusnurmath

Institute of Social Studies Trust

I must confess the contrast between Ajay and me is not only in terms of his very technical approach and my very definitely nontechnical approach. There is another difference as well. To put it bluntly, we are poles apart—both in our views and in how we have approached the paper. I am approaching it from an entirely different perspective, partly because it has been many years since I passed out of Delhi School of Economics (and this is no reflection on Professor Bardhan here from whom I learnt international economics), but also because my 11 years in Reserve Bank of India and subsequent almost 18 years in journalism have taught me that a purely econometric approach to many real-world problems does not work. In fact, I think what the financial crisis has taught us is that if you approach economics as a physical science, it does not work. It is a social science. So my approach is to look at the graphs and equations that many modern-day economists are fond of, as no more than tools that help improve our understanding. One must remember that we are dealing with human beings who are often irrational. One human being is very different from another human being; indeed the same human being also changes, minute to minute.

The issue is: did capital controls work? Ajay was very emphatic they did not work. To my mind they did work for the simple reason that the Indian economy came through the crisis without too much upheaval. Why do I say that? Because for a country like ours with a very large number of people living below the poverty line, what matters is that we were able to cushion them from the shocks of the external economy. That, to my mind, is a measure of success. So when it comes to capital controls, your equations may show you something, your graphs may show you something else. And yes, I may not have an answer as to why the Negotiated Dealing System (NDS) volumes are so high, but I do know that we did escape the worst of the crisis and that clearly shows we must be doing something right. However, I am not sure everybody in this room will agree with this very practical approach of judging the success of a policy by its outcome.

With those preliminary remarks let me now address Renu's paper. It focuses on the question of whether capital regulations are effective in insulating the domestic economy; the financial as well as the real sector, from foreign shocks. But first, why do countries want to insulate their domestic economies. Typically, there are four reasons. One, if your exchange rate is volatile, that has repercussions for your exports. Then you need to worry about hot money flows, loss of monetary autonomy, and large flows which are destabilizing in the short run, particularly if you don't have absorptive capacity like in the Indian context. Needless to say, I think that postcrisis one thing that has certainly happened—a very dogmatic view on capital controls is an absolute no, and rushing into capital account convertibility is no longer fashionable. I think even institutions like the International Monetary Fund (IMF) have realized that there may be a place for capital controls depending on each country's sensitivity and vulnerability. So certainly I think we have moved from a position where capital controls were an absolute no, no to yes, possibly, we should have capital controls.

Despite the broad endorsement of capital controls, there is very little agreement really on the kind of capital controls. Should they be the kind that we have, very convoluted? We have controls on almost every conceivable thing; we are constantly tweaking one or another aspect on both inflows and outflows. Or, should we do something like Chile and Malaysia that opted for restrictions like additional reserve requirements and quantitative restrictions on outflows.

In this context, Renu's paper is useful because she looks at capital controls in India and their effect. There have been many discussions starting from Tarapore Committee report talking about the need for capital controls and providing some kind of a blue print, but very few of these studies have been data based. Renu's study fills the gap. As Magott and Reinhart, quoted in a paper, say that the literature on capital controls has at least four various serious apples to oranges problems. One, there is no unified theoretical framework to analyze the macroeconomic consequences of control; two, there is significant heterogeneity across countries and time in the control measures implemented; three, there are multiple definitions of what constitutes success (this is where Ajay and I differ); and four, the empirical studies lack a common methodology. Furthermore, they are significantly overweighed by a couple of country cases such as Chile and Malaysia. If you look at international literature, there is not much on the Indian experience. So Renu's study really fills a gap.

However, I have a slight problem with Renu's interpretation of efficiency because I think efficiency of any instrument must be seen from the perspective of the objective and as far as capital controls in India are concerned; frankly I can give you a number of objectives, apart from the need to insulate the domestic market. Building up reserves, for instance, is a very important objective because when we opened up we had low reserves and so building up reserves was very important.

I think from my limited knowledge of econometrics, the exercise is very rigorous. The conclusion is also unexceptionable. Intuitively, one would except that capital controls would be less effective in a period of greater volatility without doing all these exercises. It is nonetheless good to have one's gut feeling reaffirmed by rigorous analysis. At the same time, one also needs to factor in whether the reduced effectiveness in the second period is partly because of the steady opening up of our controls. Ajay says we are one of the most closed economies but subsequently when he talked about the rupee exchange dollar rate, he talked about much greater volatility. Clearly these two statements are inconsistent. You cannot have a completely closed economy which did not show much change and at the same time talk of much greater volatility. The reality is we were opening up in the second period, which is why controls were less effective.

Let me now come to some specific points, where I have some reservations. As evidence of capital controls being partly circumvented, Renu talks about the overuse of participatory notes. I am not clear what she means by saying these participatory notes (PNs) were issued to offshore investors, wishing to take positions on the Indian equity and debt market beyond their regulatory limit. To my mind, PNs were issued because the registration procedure is cumbersome and also because the money that was coming through PNs was not quite kosher. A lot of it, my fellow journalists and I are convinced, though we have no proof, was Indian money that had escaped the tax net. Holders of PNs did not want to declare the beneficial owners. This was subterfuge and not really a capital controls' issue because limits on FII were fixed regardless whether money came in through PN or not. I don't think registration per se can be looked at as a capital control. Most countries have some registration requirements, thanks to the greater worries about terror financing and awareness brought about by the OECD's Financial Action Task Force (FATF). Yes, maybe our controls were more cumbersome, but that is not the point.

Again, using the IIP data to assess the impact on the real economy seems a bit flawed given the fact that manufacturing industry accounts for a very small share of our GDP. I think GDP would be a better measure. Yes, I agree there is a problem since you don't get GDP data except on a quarterly basis. Even so I am not sure that industrial production is the right indicator to take as a measure of rural activity, for instance.

While ranking financial markets according to their degree of sensitivity to foreign capital flows, the paper has placed the forex markets at No. 4. However the impact results show that the full period shock to capital flows explains only 25 percent of the stock market volatility and about 60 percent of the Rupee–Dollar volatility. The paper says "it unsurprisingly explained 60%," If that is so, I wonder why it has been ordered No. 4. Should it not rank higher?

Another suggestion is that within the overall portfolio flows one could perhaps look at whether equity flows are affected differently from debt flows and also whether capital controls have affected FDI. Have they affected FDI differently than the portfolio investments of FII? That leads to the related question of China's capital controls, how would one measure their success given that China has received much more FDI rather than FII, whereas we have much more FII vis-à-vis FDI? In fact there was only one year when we have had the happy position where FDI was more than FII, but that again unfortunately has reversed.

My last point—this is actually more a rejoinder to Ajay. He spoke of every other country being much more open than us. I am not sure that there is a great virtue in following the herd because the crisis has shown that many countries that rushed in and did whatever was fashionable are regretting their actions today. So maybe it makes sense to take into account country-specific factors, for instance, the problems peculiar to a country as multidimensional and multifaceted as India. We have a vast number of poor, each state is very different. So maybe there is no great virtue in following the herd. Rather, you need to do things your own way.

Critics often say that India wants to do everything its own way and we like to say we are different. But to me the litmus test is really that we have come through relatively unscathed. I say relatively because maybe financial markets are not happy, but yes, I think that the average Indian is certainly better off in today's circumstances than he would have been had we been like an Indonesia in the 1998 crisis. Let me end with a quote from Keynes, now that Keynes is back in fashion. He said, "Above all let finance be primarily national," and I think that is something that we still need to think about.

General Discussion

T.N. Srinivasan sympathized with Ajay Shah's belief that the controls were largely ineffective. He argued that it was not possible to conclude that the financial controls were responsible for India's economy being relatively unscathed by the global financial crisis because the paper did not provide a model that integrated the financial and real sectors. Thus, there was no basis for determining outcomes in the absence of the controls. Abhijit Banerjee also spoke about the importance of a more developed theoretical model. He questioned the definition of "big" or "small" frictions and wondered how much volatility could be expected under normal conditions, and he thought that a theoretical model could help in clarifying those issues.

Rajnish Mehra also agreed about the need for a stronger theoretical model. He questioned the reliance on UIP because it was known to fail even in the absence of controls. He thought it would be more effective to focus on the NDF market and covered interest parity. It might be possible to determine if there was an arbitrageable region. He also thought that the huge size of the 2008 shock would dominate the regression analysis. Finally, he pointed out the response of the bond market to a foreign shock should be muted relative to that of the equity market even in the absence of controls. Equity yields are very sensitive to growth rates, whereas the bond rate would only change in situations that altered expectations of default.

Ashok Lahiri expressed concern that capital controls could be easily circumvented. He also thought that, while the controls might reduce volatility, they would also reduce investment in India. How should that trade-off be evaluated?